Stephen W. Hawking is the ~~Luca~~ ...atics at Cambridge University, a position formerly held by Isaac Newton. He is famous for work on singularities in spacetime, like the big bang at the creation of the universe. With Roger Penrose he showed that such singularities must occur inside black holes. He has also given theoretical physics "Hawking radiation" from black holes, a first piece in the puzzling connection between quantum mechanics and general relativity. His books include *A Brief History of Time*, *Black Holes and Baby Universes and Other Essays*, and *The Universe in a Nutshell*.

Igor Novikov is the leader of the Theoretical Astrophysics Center of NORDITA, the Nordic Institute for Theoretical Physics, in Copenhagen, and is professor at the Copenhagen University Observatory. He is one of the world's foremost researchers in relativistic astrophysics, and in the possibility of time travel. Among his books are *Black Holes and the Universe* and *The River of Time*.

Kip S. Thorne is the Feynman Professor of Theoretical Physics at Caltech. He has been responsible for many advances in our understanding of the role of curved spacetime gravity in astrophysics. Along with Ron Drever and Ray Weiss, he founded the LIGO project to build a gravitational wave detector, and he has inspired worldwide activity in building such detectors. He is the author of many books, including the popular *Black Holes and Time Warps*.

Timothy Ferris is the author of *The Whole Shebang*, *Coming of Age in the Milky Way*, and *The Mind's Sky*, and is a two-time winner of the American Institute of Physics Prize for science writing. His work has appeared in *Harper's*, *The New Yorker*, *The*

Nation, and *The New Republic*. He is professor emeritus at the University of California at Berkeley.

Alan Lightman is an acclaimed novelist and the John E. Burchard Professor of Humanities and a senior lecturer in physics at the Massachusetts Institute of Technology. His novels are *Einstein's Dreams*, *Good Benito*, and *The Diagnosis*. He has published his essays, stories, and reviews in *The Atlantic*, *Granta*, *Harper's*, *The New Yorker*, and the *New York Review of Books*.

Richard Price is a professor of theoretical physics at the University of Utah; his research has focused on spacetime physics, and he made early breakthroughs in black hole physics. He has also worked on problems in microelectromechanical devices, and issues in science education. He has coauthored *Problem Book in Relativity and Gravitation* and *Black Holes: The Membrane Paradigm*.

The Future of
SPACETIME

Stephen W. Hawking

Kip S. Thorne

Igor Novikov

Timothy Ferris

Alan Lightman

Introduction by Richard Price

W. W. NORTON & COMPANY
New York • London

Composition by Tom Ernst
Manufacturing by the Haddon Craftsmen, Inc.
Book design by Charlotte Staub

Library of Congress Cataloging-in-Publication Data
The future of spacetime / Stephen W. Hawking . . . [et al.] ; introduction by Richard Price.
      p. cm.
    Includes bibliographical references and index.
    ISBN 0-393-02022-3
    1. Space and time. I. Hawkings, S. W. (Stephen W.)

    QC173.59.S65 F87 2002
    530.11—dc21   2001052105
ISBN 0-393-32446-X pbk.

W. W. Norton & Company, Inc., 500 Fifth Avenue, New York, N.Y. 10110
www.wwnorton.com

W. W. Norton & Company Ltd., Castle House, 75/76 Wells Street, London W1T 3QT

1 2 3 4 5 6 7 8 9 0

CONTENTS

PREFACE

This book is not pure. For one thing, it is an amalgam of curiously different essays that have been put together with the weld marks showing. Our editor at W. W. Norton, Ed Barber, has been very supportive throughout the welding, but several times we thought we heard him mutter the word "hodgepodge" nervously (but supportively). The book, in fact, is just the way it should be, an eclectic collection that is charmingly heterogeneous. This is the way science and scientists are—mixtures, not tightly constrained and often not very obviously organized.

There *are* organizing principles behind the book: The contributions are excellent and are readable—there's hardly an equation. They are all related to the modern physics of space and time. Most importantly, they are adaptations of popular talks given on June 3, 2000, to honor the sixtieth birthday of the California Institute of Technology's Kip Thorne. But we concede that the content creates some quaint juxtapositions. There are three essays that communicate science, one essay about communicating science, and one essay about the difference between science and communicating.

This impure book was born in a deception. It is traditional to have a celebration to honor important scientists on their sixtieth birthday. Kip Thorne is not only an important scientist but important to each of us personally. We wanted to make this occasion something really special, but his modesty was an inconvenience. We therefore resorted to what might be called misrepresentation, or even fraud, in getting Kip to agree to be honored, and to take part. The ducks were already all lined up before Kip heard any quacking. Five famous speakers had agreed to give their talks; Caltech's public events center, Beckman Auditorium, had been reserved. By the time Kip knew the whole truth, it was too late for him to back out.

We present here adaptations of the talks given at Caltech on that day. The speakers invited were distinguished and accomplished, yet would draw a crowd. It says something about Kip's place in the community that there were no second choices. All those who were asked to speak accepted the invitation. They accepted no fee for their talks, or for the use of adaptations of their talks in this collection. On Saturday, June 3, 2000, the talks were free. The royalties from this book will go to a scholarship fund at Caltech attached to Kip's name.

What is interesting enough to draw a crowd to an auditorium should also be of interest to many of you who could not be in the right place at the right time. This volume may lack the immediacy of the physical presence of the speakers, but it allows for more deliberate chewing of some gourmet dishes that should not be rushed.

In one essay to follow, Igor Novikov, director of the Theoretical Astrophysics Center in Denmark's NORDITA (Nordic Institute for Theoretical Physics), tells us about time travel—a subject that seems strange and wonderful even to a scientific community in which black holes are as comfortable as old shoes. We are introduced to the subject with simple explanations and simple mechanical models of how to avoid paradoxes when traveling back to an earlier time. Even if nonparadoxical, time travel may be impossible. Stephen Hawking, Lucasian Professor of Mathematics at Cambridge and one of the world's most famous scientists, gives the results of his investigations of the question "just how impossible." We learn that this question requires going to the very edge of what is understood about physics, and that its answer will require going a bit further. Kip Thorne, in his essay, tries to travel in time to the future. (We all do this eventually, but Kip goes out ahead on a scouting mission.) Gravitational wave astronomy will be a reality in the near future, and Kip shares his enthusiasm for the exciting discoveries that it will produce in the not-so-near future.

The last two essays form a set somewhat distinct from the scientific explanations of the others. One is by Timothy Ferris, an outstanding science writer and journalist, who set the standards, and set them high, for explaining astronomy and cosmology, with books like *The Red Limit, The Whole Shebang,* and *Coming of Age in the Milky Way.* He tells us about the need and the difficulty of explaining science, and includes a portion of a screenplay that falls somewhere between science and art, or in both. Alan Lightman certainly lives in both these worlds. A leading physicist with a passion for writing, he has become a leading writer with a passion for physics. For those not in physics or in the MIT Writing Program, Alan is probably known best for the 1993 bestseller *Einstein's Dreams.* Having experienced the very different kinds of creativity that are part of science and part of art, he has the rare authority to contribute his essay comparing the two.

In addition to adaptations of the five talks, a very brief introduction to the physical ideas about spacetime, and of the history of these ideas, is provided by Richard Price, a theorist in the

Physics Department of the University of Utah. This introduction
sets the stage for the science revealed, chronicled, and pondered
by Timothy Ferris, Stephen Hawking, Alan Lightman, Igor
Novikov, and Kip Thorne.

Acknowledgments

This book owes its existence to the Kipfest celebration at
Caltech at the beginning of June 2000. It is therefore in debt to the
many people who helped bring about that event. We seven were
the organizing committee for that event, but were only a small sub-
set of those who contributed. Some of these must be mentioned.

The book and event could certainly not have happened without support, both financial and logistical, from the Caltech
administration. In particular, thanks go to David Baltimore, president of Caltech, and Thomas Tombrello, chair of Caltech's
Division of Physics, Mathematics and Astronomy. David
Goodstein, vice provost of Caltech, deserves special thanks for
acting as the master of ceremonies for the talks.

In addition to financial support from Caltech, Kipfest received
much needed funding from David Lee and his company, Global
Crossing. Thank you, David.

There were many who helped in many ways: the staff of
Beckman Auditorium, the Caltech public relations people,
Caltech's Athenaeum, Lynda Williams (the "Physics Chanteuse"),
and all those who traveled from across the world or across
Caltech's campus to take part. Thank you all!

Eanna Flanagan   *Clifford Will*
Sandor Kovacs   *Leslie Will*
Richard Price   *Elizabeth Wood*
Bernard Schutz

The Future of SPACETIME

INTRODUCTION:
Welcome to Spacetime

Richard Price

A Happening in Spacetime

It's funny how long you wait to ask some of the most important questions, even questions about your own life. You may
have to wait for an event that encourages stepping away and
looking back. An event of this sort was a celebration of the sixtieth birthday of Kip S. Thorne. Kip (he is uncomfortable with any
stuffier title) is one of the country's best-known theorists on the
physics of spacetime, and a great popularizer of his strange science. He is also a person whose humanity is as singular as his
intellect, a person who has affected the lives of many who have
worked with him. A sixtieth birthday symposium for an important scientist is something of a tradition in physics, but there was
clearly a more complex feeling in the air at Caltech in June 2000.
Attending the celebration was as much an outpouring of affec-
tion as of homage.

Duty and desire drew physicists spanning Kip's career, from
the midsixties to the present, so that a student of spacetime
physics walking past Caltech's Ramo Auditorium during coffee
breaks on June 2 and June 3 would see a living museum of the

science of the era. The exhibits in this museum included colleagues who had forgiven each other some past slight and were
again talking, physicists who were awkwardly introducing new
spouses, and colleagues, once student and teacher, who were
now brought to a more relaxed equality by the passage of time.
The fact that the gathering coincided with the advent of a new
millennium might have impressed a more impressionable
crowd, but for this one there was a more appropriate portent: the
imminent completion of a worldwide system of detectors to map
out gravitational waves, oscillations in spacetime.

The birthday was a reminder of the passage of time. The gathering of old friends and rivals created the monochrome feeling
of a Swedish movie despite the palm tree surroundings. It was a
time to ask some deferred questions, such as: What drives reasonable people (an assumption is being made here) to spend
their lives studying the nature of space and time?

This introduction is being written at the end of a century
characterized by science, especially by physics. It was Albert
Einstein, after all, who was chosen by (the ironically named)
Time magazine as the person of the century. Einstein had started
the century off impressively in his miraculous year 1905. In that
single year, he gave statistical proof of the atomic nature of matter, and with his Nobel Prize–winning explanation of photons
hitting metal surfaces, he helped to push forward the quantum
revolution that he never felt comfortable with. But for scientists
as well as nonscientists, the response to "Einstein" in a word
association test would be to cite his third miracle of 1905: "relativity," the theory of the structure of space and time.

What was it about that work, and not his more relevant, "useful" work on atoms and photons, that made Einstein a celebrity
and hero? It is probably the fact that we deal daily with spacetime, and we think we know it. Atoms are too small, photons are

too many; we don't have strong opinions about such things. When we are told news about them, we accept it as the orderly progress of science. Matter is made of (sort of) indivisible units; light has both wave and particle nature. The nonscientist has no evidence for contradicting the first statement, and no clear understanding of what is meant by the second. But in 1905 Einstein also told us that time is not a universal clock ticking at the same rate for everyone, and that one twin who goes off on a high-speed rocket trip will not age as much as the other twin who stays at home. *This* is understood and is outrageous. What caught the popular imagination was that it is impossible, and yet it is correct.

We are fascinated with being wrong. It teaches us about ourselves. Not only are there things we don't know, but the things we do know can be wrong.

Relativity, or spacetime physics, with its aura of black holes and an expanding universe, grabs our attention because it is the stuff of daily life—space and time—made exotic, as if the librarian has driven by in a Ferrari wearing a sarong. This explains, I think, the reason for the enduring public fascination by scientifically literate nonprofessionals. It also explains relativity's importance to those with too little patience and perhaps too much self-confidence. Every relativist has the experience of receiving, a few times a year, a new theory of relativity from a technically inclined nontraditional thinker who hasn't read "all the books" but knows where Einstein went wrong.

An answer to the "why" question is not so clear for those of us who are Kip's students, colleagues, and coworkers. We have read "all the books" and we work with details. My own research, for example, deals mostly with applied mathematics, the same applied mathematics that could be directed to honest work like hydrodynamics or chemical engineering. The wonder that these efforts are directed to collisions of black holes is easily lost to familiarity. The same thing happens on airplanes. Squeezed into our seats we grumble about minutiae and show no awe that we have left the surface of Earth. But sometimes, watching from a hill, we see a huge jet slide silently above a city and we are stunned. In the same way, I sometimes look up from the computations and remember I am trying to unravel puzzles about regions of the universe from which no escape is possible; this is the stuff of my daily work! (And stranger still, I get paid for it.)

The contributions in this volume illustrate several different themes that are part of a larger theme. Stephen Hawking and Igor Novikov tell us about time travel—a subject that seems strange and iconoclastic even to a scientific community in which black holes are as comfortable as old shoes. Kip Thorne then goes in the very different direction of gravitational waves, the spacetime oscillations that will be detected in the near future by worldwide experiments, and Kip makes predictions about what will be discovered. Whereas the time-travel musings deal with what the laws of nature could make impossible, Kip muses about what technology will make possible. Very different kinds of musings appear in other chapters. Alan Lightman gives his insights on the differences between the creative acts of writing and of solving a scientific problem. How can the wonder of this kind of science be communicated to those without the technical background? Tim Ferris, who has been exceptionally successful at doing precisely this, gives us some answers in his contribution.

This introduction is meant to set the stage on which the contributors will perform. I will present a minimalist sketch of just what it is that physicists do when they say they are working on spacetime. Since my introduction would inevitably lose in a contest of comparison with certain other introductions to the subject, let me lower the reader's expectations. Here I will not describe the interaction between the technical and human sides of the story, the way Kip does so beautifully in his recent popular book *Black Holes and Time Warps*.[1] I will also not be explaining the introductory ideas with as much completeness and mathematical clarity as Edwin Taylor and John Wheeler do in their wonderful little book *Spacetime Physics*.[2] Should the reader's interest be aroused by this introduction, those are excellent books to turn to. Here I will be touching the surface only, and sometimes floating just above the surface. The main goal of this introduction is brevity, and that has been achieved. I hope it is not the only achievement. I do think it gives some substance to some of the ideas that arise in the contributions compiled here.

The questions are not new. An interest, perhaps an obsession, with the nature of space and time is as old as human thought. The classical thinkers had much to say about the subject.[3] Some of it now seems quaintly naive, some of it is still impressively deep. (To me, Zeno in particular seems to have aged well.) The discussion here will be limited to modern ideas, ideas that took thousands of years to evolve and that find their precise implementation in mathematics. It is a pleasant surprise that the issues of such a modern discussion are accessible to those without an extensive background in mathematics and physics. What will be important is to deal at the outset with certain central words that come from everyday discourse but have taken on a special and precise meaning in connection with spacetime. Physics is not very different from other endeavors in this sense. If you don't know what it means to "fold" an egg, you won't be able to cook an omelette; if you don't know what an "event" is, you can't understand the geometry of spacetime.

Disagreeable Observers

An introduction to the special words doesn't have to be abstract or uncomfortable. The proof of this is the book by

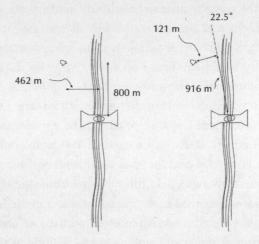

Figure 1. The same location with different coordinates.

Taylor and Wheeler, which introduces the words, the ideas, and the mathematics so clearly. I will pay the sincerest form of flattery by taking, with modification, some of the images from the beginning of that book.

The story opens with a person standing at the middle of a small bridge crossing a straight, narrow river running through a flat landscape. She is facing directly upriver and wants to give a *quantitative* description of the location of places of interest, like the bell on the old church. There are all sorts of ways she could do it. She could say that the bell is 924 meters away from her, and the direction to it is at an angle of 30 degrees toward the left. Alternatively, she could note that the bell is 800 meters "forward" (in the upriver direction) and 462 meters "left" (meaning 462 meters to the left of the river—see Figure 1). What is common to both methods of description (and to any other method) is that she must specify *two* numbers. For that reason we say that the set of locations on the landscape is a two-dimensional world. Often in physics, measurements are said to be made by an "observer" and the method of locating points is a "frame of reference" associated with the observer. The particular numbers our observer arrives at (such as 800 meters and 462 meters) are called the "coordinates" of a location.

The existence and importance of these special words correctly suggests that there can be other observers and other frames of reference. This is, in fact, what relativity is all about: the relationship of measurements (that is, coordinates) in different reference frames. It is crucial, then, that we have another observer and that our observers disagree about measurements. Let our second observer be standing in the middle of the bridge, right next to the first observer. He also is compiling a quantitative description of the landscape, the two-dimensional world of locations, and he also does it using the "forward and left" method. This would not produce any insights if he were facing upriver: he would agree with the first observer, and no instructive disagreement would follow. We will therefore have him face in a direction different from that of the first observer. He will face one-quarter of the way between her "forward" and "left" directions, or equivalently he will be facing 22.5 degrees left of the upriver direction. This gives him a different reference frame and puts him facing more nearly in the direction of the old church bell. As a consequence he measures different coordinates: the bell is 916 meters forward and 121 meters to the left.

We understand that there is really only one bell in only one location, and that the two observers do not disagree on the location, only on the numbers (coordinates) that characterize it. We know furthermore that there must be a relationship between the coordinates marked by the first and second observers. This relationship between coordinates of any of the observers is called a "transformation," another one of those special words. It is a mathematical expression of a kind of relativity, a relationship of coordinates measured by one observer and by another. That relationship is given by formulas that would be taught in high school mathematics. These are not difficult formulas, but lines of mathematics distract attention from the basic ideas that they represent. Rather than present formulas, I show, in Figure 2, the

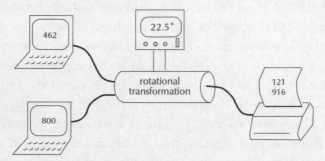

Figure 2. The formulas of the rotational transformation
pictured as a machine.

mathematics as a device that takes as input the left and forward locations according to the first observer and gives as output the values as marked by the second observer. Of course, there must be another kind of input to this machine: the way in which we specify the relationship of the two reference frames. In our case, this means that we must enter the specification "22.5 degrees."

The machine in Figure 2 is really just the formulas, the formulas that, taken together, a high school mathematics teacher might call a "rotational transformation." The machine could actually be a *very* simple computer chip designed to do nothing more than perform the simple computations of the rotational transformation. In the wonderful age we live in, it would cost pennies to build each of these machines.

There is an important feature of the system of reference frames that our observers have used. To see this best, consider a very different way of identifying structures: by tax ID numbers. Suppose that all structures in town are identified by numbers for purposes of imposing and collecting property taxes. These numbers were assigned in some manner for the convenience of accounting. Let us go on to suppose that the system for assigning tax ID numbers changed (perhaps a new computer was purchased for the accounting office). A table must be kept of the relationship between the old ID numbers and the new; two items in that table are shown here.

BUILDING	OLD ID	NEW ID
bell	50070	CX23-004
barn	34210a	BX48-213

This table is a relationship between two systems of coordinates, and—in a sense—is a kind of transformation.

Intuition screams out that this kind of transformation is somehow very different from the rotational transformation, but just what is the real difference? Our intuition insists that the tax IDs and their relationship are arbitrary; they are assigned for the convenience of some computer. Any system of identification could be used and could be "transformed" to any new system of identification. On the other hand, the forward and left distances measured by the observers on the bridge are not arbitrary. But what institution, what higher authority, prevents them from being arbitrary? What guarantees that the rotational transformation is correct? Ultimately it is the fact that the setting on which the structures are located has some geometry, the geometry (usually called "Euclidean geometry") of a plane. Any two locations, say the bell and the barn, on that plane have a certain distance between them. That distance is an immutable truth, though the coordinates (the leftness and forwardness) used to describe location are variable. Because there is an immutable truth not subject to the vagaries of reference frames, the relationship between distance-based coordinates in different reference frames cannot be "just anything."

Galilean Relativity

Armed with enough special jargon we can now dip a tentative toe into spacetime. Just as locations are the places on a landscape, "events" are the places in spacetime. An event is a certain place and a certain time. It is a location in time as well as in space. Clearly the world of such events—the world we call spacetime—is four-dimensional. It takes three coordinates to specify the "where" of an event, and one coordinate to specify the "when."

To get a more specific understanding of this, we will need to look at the disagreement between observers, and hence we will need at least two different observers, with two different reference frames. One of our observers will be a farmer standing in her field while a railroad train slowly goes by, and the second observer will be the train, or a passenger sitting on the train. Our observers on the bridge found a disagreement in the coordinates they used because they faced in different directions and hence had different meanings for "forward" and "left." We are interested in a different kind of relationship now, so we avoid the forward/left disagreement by having our farmer and passenger facing in the same direction. To generate disagreement we will have the train move in the forward direction at 3 m/s. (From this point on we will sometimes abbreviate meters, kilometers, and seconds as m, km, and s, respectively; thus, 3 m/s means a speed of three meters per second.)

There are two important features of the scenario we are constructing. One is that we have simplified the action by having only one spatial dimension be important. Only locations along the direction of the track are interesting; distances perpendicular to the track are trivial because all events take place right along the track. We have therefore eliminated two coordinates. More important, we have added one coordinate. By introducing motion into the story, we have opened the door for the time coordinate to enter.

For simplicity we will say that the time $t = 0$ is the moment when the passenger is just passing the farmer. It is the moment when they would agree about observations of an event. It is crucial for us to witness disagreement about events, so we will create an interesting incident at time $t = 2$ s. Suppose a hawk snatches a mouse in that incident, and that it happens at a location 16 m forward of the farmer. Since the passenger (moving at 3 m/s) has moved 6 meters past the farmer by the time $t = 2$ seconds, the mouse snatching will happen at a location only 10 meters forward of the passenger. This simple situation is pictured in Figure 3.

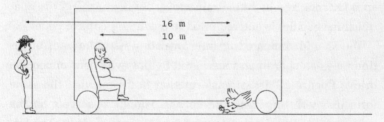

Figure 3. An event observed in two different reference frames.

The farmer and the train constitute two different reference frames in which the location and time coordinates of events are marked. The relationship between the coordinates is, of course, a transformation, and this one is known as the "Galilean transformation." The general idea of the relationship it gives to coordinates measured in moving reference frames is called "Galilean relativity."

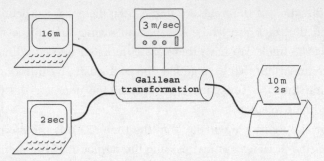

Figure 4. The Galilean transformation between reference frames
that are moving with respect to each other.

In Figure 4 the mathematics of the Galilean transformation
(actually two very simple equations) is pictured as a machine.
The spacetime locations measured by the farmer are input into
the terminals, and the locations in the train reference system are
the output. Of course, the machine must be set for the right relationship between the reference frames. That is given as the
speed, 3 m/s, at which the train moves by the farmer. This relationship specifier is entered into the machine's control panel.

You should compare this transformation with the transformation between coordinates measured by the two observers on the
bridge (Figure 2). This transformation has somewhat the same
form but with important differences. For the observers on the
bridge, there was a complete mixing of forward and left. The forward and left of the first observer determined the forward and
left of the second observer. In our transformation between train
and farmer, there is only a partial mixing. The distance coordinate measured by the farmer is affected by both the distance and
time coordinates measured by the passenger, but the time measured by the farmer is not influenced by the distance coordinate
of the event. The second formula built into our transformation
machine simply says that 2 seconds equals 2 seconds. Space
does not mix into time.

Actually, what it tells us seems obvious, too obvious to deserve a fancy name. The expression "Galilean relativity" is, in fact, a modern expression that we only use *now* in order to distinguish the obvious classical notions of space and time from the anything-but-obvious notions in Einsteinian relativity. These notions, along with aesthetics and psychology, are the background for the development of scientific revolutions. In many ways it is a story of the evolution of what is and is not obvious.

Though Galilean relativity was obvious, it was of great importance to the physics of Newton. Newton had given the world the commandment $F = m \times a$; acceleration is proportional to force. All observers (farmers, trains, and so forth) see the same sources of force. They see the same bending of a bow, for example, and hence the same force imparted to an arrow. When the bowstring is released, they must see the arrow undergo the same acceleration. If they did not, then Newton's commandment might work only in some reference frames; it could not work in all. But acceleration is a measure of the way spatial location changes as time changes. The comparison of acceleration in two different reference frames depends on the transformation rules between those reference frames. It turns out that a consequence of the Galilean transformation is that acceleration *is* the same in all reference frames. The farmer and the passenger on the train, when marking the locations of the arrow at a sequence of times, *do* arrive at the same number for the acceleration of the arrow while the bowstring is pushing on it. Newton's towering commandment *does* work in all reference frames.

This bothered Newton! For his own philosophical, psychological, or aesthetic reasons, he yearned for a physical universe in which a special reference frame—the farmer's reference frame, perhaps—was the one and true reference frame for physics. But there were no reasons in his physics for believing in the existence of such a "correct" frame of reference. Perhaps Newton's attitude was rooted in a human need for there to be something solid and absolute, one true frame. If so, it is interesting that the philosophical or psychological fashion of physicists has changed so much. To modern eyes and minds, the democracy of reference frames is an attractive feature of Newtonian mechanics.

Maxwell Creates a Crisis

Aside from Newton's unease, the physical world seemed to make sense during the eighteenth and nineteenth centuries. Understanding doesn't move forward smoothly like a skater on ice. It moves forward in a start, like a stuck cap being pried from a jar. The longer the jar is closed, the more tightly bound the lid becomes. An idea like that of an Earth-centered universe was stuck so tight because it had been in place for such a long time. For the centuries of pre-Copernican astronomers there was no question whether the Earth was the center of the world. If difficulties arose, they would look elsewhere for remedies. Those astronomers constructed an extraordinarily complex calculational method to predict and explain the motion of heavenly bodies. An originally simple method of prediction was found to be inadequate when observations of planetary motion improved. Mathematical constructions, "epicycles," were invoked to improve the predictions, and the basic theory was coerced into an appearance of working. This cycle of improvements continued, first in adding astronomical observations, then in adding more unwieldy features to the method.

When we look back at what they were doing, we are incredulous. How could they not see that the simple elegant idea of a Sun-centered world explained everything? They had not so much missed what now seems obvious, as they had been seduced, step by step, down the wrong path. The beginning of the path pointed in a reasonable direction, and from well along the path it was hard to see that there were alternative paths. There are critical moments in science when pressure has built up to the point that the lid is about to come loose, but these special moments require special minds. Facts and misconceptions are taught together. Those who knew the details of planetary motion had been steeped in Earth-centered discourse. At the special moments a genius (surely a justified term) is needed who is able to keep separate what is robust fact and what is fragile dogma. The critical moment for understanding the solar system came in the late sixteenth century, and it was Copernicus who was the genius at the right place and time.

A critical moment arose at the beginning of the twentieth century, and at this right time and right place, it was Albert Einstein who was the right person. It is a common fantasy to go back in time with a crucial fact or bit of understanding, something widely known now but unknown in the past. The right material to use for the filament of an electric lightbulb, the insight that nebulae are distant islands of stars, the nuclear model of the atom—at the right moment in the past, each of these bits of magic would qualify the holder as a genius. The magic that was needed at the beginning of the twentieth century was a simple insight that is more or less understood by most undergraduate physics majors today. This magic is mathematically simple, can be succinctly stated, and was rooted in the work of another genius.

At the end of the nineteenth century, James Clerk Maxwell added a missing piece to an equation called "Ampère's law," and in so doing tied up all that was known about electromagnetism (electricity and magnetism). To those who practice it, mathematics can have a beauty. Maxwell's theory not only explained all that was known about electromagnetism; it did it in such a beautiful way that it is still a role model for other theories.

Maxwell's theory consists of four equations. These equations deal with electric and magnetic forces, but they also involve space and time. A typical term in one of Maxwell's equations would multiply a magnetic force at some point in space by the spatial coordinate of that point. Another typical term represents the rate at which the electric force is changing as time changes. Maxwell's equations are relations among such terms: the first term plus 4π times the second term is equal to the third term, something as simple as $A + B = C$. Suppose our farmer, standing by the railroad tracks, were to make a calculation of each of the terms in one of the Maxwell equations. She would multiply the magnetic force at a point by the spatial coordinate of that point in her reference frame; she would find how much the electric force changes in every millisecond of her time; and so forth. She would then see whether the terms she has calculated all "add up," whether they satisfy the appropriate Maxwell equation.

Suppose she finds that the terms do add up, that for her the Maxwell equation is correct. Our curiosity then turns to the railroad passenger. The way he assigns coordinate location to events will be different from those of the farmer, so the values of the terms in the Maxwell equation will be different from the values the farmer calculates. We must then ask a crucial question. Can the Maxwell terms "add up" for the passenger as well as for the farmer? Do all the changes in the terms conspire so that Maxwell's theory works for both observers?

The answer is no. If we relate the terms with Galilean relativity, Maxwell's equations cannot work for both farmer and passenger. They can be valid only in a single reference frame. Newton's theory of forces and motion works in any reference frame; Maxwell's theory of electromagnetism can work in only one. As the nineteenth century ended, it seemed that Newton's suspicion two centuries earlier had been correct. There *was* a special reference frame for physical law; it was the reference frame in which

Maxwell's equations worked. Who could argue against calling this the true reference frame of the physical world?

So, just what was this true frame? The experiments to find the true frame required great precision and were not easy. The experimental search is an oft-told tale, and the end of the story is well known. No special frame was found. Maxwell's theory worked for both farmer and train. The experiments said this was the case, but the mathematics said that this was impossible. That impossibility, of course, was based on a certain way of relating the spatial coordinates and the time of the farmer and the train. That is, the impossibility was based on Galilean relativity. The scientific (though fictional) detective Sherlock Holmes told Dr. Watson that when confronted by a mystery, one must reject the most implausible alternatives until only one possibility remains, and that possibility—no matter how implausible—must be the answer. But prioritizing plausibility is a subjective matter. For most scientists, the only possible conclusion was to modify Maxwell's beautiful theory. Something like epicycles had to be added. These modifications were awkward, but even worse, no modifications could be found that worked. They all contradicted experimental evidence.

Einstein's Revolution

Albert Einstein, a patent clerk in Bern, Switzerland, had another set of priorities. To him it was plausible that Galilean relativity was not correct. To him it was conceivable that the location and time coordinates of the farmer, and of the passenger, were related in a different way than by Galilean relativity. Another relationship of coordinates was needed. It is ironic that the new relationship had already been worked out by the Dutch physicist Hendrik Lorentz. That relationship is now called the Lorentz transformation of coordinates, not the Einstein transfor-

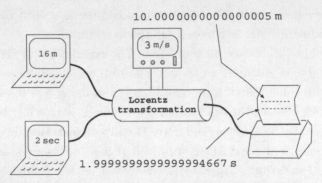

Figure 5. The Lorentz transformation between reference frames
that are moving with respect to each other.

mation. For the death of the mouse, the event previously
described with the Galilean transformation, the Lorentz transformation is illustrated in Figure 5.

The difference between the numerical results and those for the
Galilean transformation (in Figure 4) are tiny, but this is because
the control panel of the transformation is set at 3 m/s. If we
increase this setting, the differences increase. Thus if the train is
moving by the farmer at extremely high speed, the differences can
be substantial. In the details of the way in which the Lorentz
transformation depends on this speed setting, there is something
quaint. If we increase the speed to something very near 300,000
km/s, then the transformation starts making pretty exotic predictions, and in fact we cannot chose a speed greater than 300,000
km/s. (In the mathematics, it turns out that this would involve
taking the square root of a negative number.) This upper limit on
the relative speed that can be handled by the Lorentz transformation is so special to the transformation that we give it a special
symbol. This 300,000 km/s speed is denoted c. (It is actually
299,792.458 km/s, but we will round it slightly.)[4]

Lorentz believed that the "time" and "distance" described by his mathematics were not *true* time and distance, but rather the time and distance that would be measured by instruments unavoidably distorted by electromagnetic effects. If the farmer represented the frame of absolute rest in the physical universe, then the train passenger would necessarily be moving, and—according to Lorentz— the material in his measuring instruments would be affected by electromagnetic fields in such a way that they would give false readings of space and time. If these false distances and times were used in Maxwell's equations, the equations would seem to work. Thus, the deck was stacked so that Maxwell's equations always *seemed* to work. When we look back at this now, it's hard not to think "epicycles," but we should remember that the wrong nature of distance and time were so very obvious, and the lid had been on the jar for such a very long time.

Einstein pried the lid off the jar by telling the world that the Lorentz transformation was not a description of distorted measurements but of actual distance and time. It was not something special to the theory of electromagnetism, but something basic about the nature of the physical world. The Lorentz transformation, what Einstein would have replace Galilean relativity, is a simple set of equations. The equations are at the level of simple high school mathematics and contain nothing more sophisticated than a square root operation. This is why Einstein's bit of magic seems to have had a different flavor than previous conceptual breakthroughs. It was—in a sense—so damned easy, so lacking in complexity. Copernicus had to spend years with solar system observations, and Newton had to invent the calculus to demonstrate the application of his laws of motion. Einstein had only to point to a set of simple equations and tell the world to think of them another way.

It was the enormity of the conceptual leap, not the complexity of its context, that spoke of Einstein's genius. Other jumps forward had required revolutionary changes in a view of the world: the Sun, not Earth, was the center of the solar system. But in these changes we were replacing knowledge that had been learned. Einstein's revolution required us to abandon what our eyes, heads, and hearts knew to be true.

Spacetime Diagrams

The mixing of one distance measurement and one time measurement looks *something* like the way in which one of the observers of Figure 1 mixes together the two kinds of distance— "forward" and "left"—to get new values of forward and left. There is no absolute meaning of forward and left. When you turn a bit to your right, your new direction of forward mixes together your old idea of forward with some negative amount of your old idea of left. If you turn completely (that is, by 90 degrees) to your right, you make a complete interchange of the two kinds of distances. Your new left is your old forward, and your new forward is the negative of your old left.

In a sense, the relationship between two reference frames for events is similar to one reference frame being turned in spacetime with respect to the other. In a new reference frame, the time and distance of the old reference frame get mixed together into the new time and distance, just like an observer on the bridge turning. *The analogy, of course, cannot be perfect!* For the observers on the bridge, after all, forward and left were indeed the same sort of thing: both distances. We just happened to give them different names associated with the particular way in which we are facing. In the spacetime of the Lorentz transformation, on the other hand, time and space really are not the same stuff. Indeed, one of the differences is that with the Lorentz transformation we cannot completely convert time to distance, or vice versa.

What this really means is nicely illustrated with a kind of picture called a "spacetime diagram." This is a sort of map of the location of events in spacetime. On this map we mark the numerical value of the locations and times of the events. The "location," or horizontal, axis uses the units of kilometers, surely a reasonable unit for distance. For the "time" (vertical) axis, however, we do something that adds a bit of complexity to the explanation, but which pays off big for the usefulness of the diagram. We denote time also in units of kilometers. To do this we simply multiply the time coordinate of the event by $c = 300,000$ km/s. Thus, if the time of an event is 1 second, we mark it down as being at a time of 300,000 kilometers. To say that an event has a time coordinate of 1 km is the same as saying that its time coordinate is 1/300,000 of a second.

Since we are marking down particular location and time coordinates, we must be using a particular reference frame. A spacetime diagram always corresponds to a particular reference frame; the location of events in a different reference frame require a different spacetime diagram.

Let's give a particular example, not too far removed from our farmer-and-train story. Suppose that the diagram in Figure 6 corresponds to the reference frame of the farmer. To make things interesting we'll suppose that our single slow train is replaced by two *extremely* fast trains on parallel tracks. Let B be the event of blowing the whistle on one of the trains, and C of blowing the whistle on the other. By a happy coincidence, the two events happened at the same time, the time the farmer calls 3 kilometers, or equivalently 1/100,000 of a second. Event A is the event at which the trains happen to be at the same place, the place the farmer is standing and that she calls "0 kilometers." A tiny amount of arithmetic tells us that event B happened on a train going at a speed of 100,000 km/s. Since the other train, which

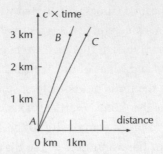

Figure 6. Spacetime diagram for the farmer's reference frame.

had the *C* event, goes 50 percent farther in the same amount of time, that train is going at 150,000 km/s.

In Figure 6 a straight line has been drawn connecting events *A* and *B*. Every point on this line could represent an event that takes place on the slower train. In a sense, it is a plot of all the events of that train's existence. This history of the train is called its "world-line." The faster train has another worldline, the line *AC*, which is tilted more toward the horizontal. This greater tilt toward the horizontal means "more space is covered in a given amount of time." That's a lengthy way of saying "faster." How tilted can a worldline be? Suppose it is tilted by 45 degrees. This means that a distance of 1 kilometer is covered for every kilometer of time (or 1/300,000 of a second). In other words, it is going at 300,000 km/s. It is going at *c!* So the most tilt a worldline can have is 45 degrees. It would violate physical law to go faster than *c*, so it is physically illegal to tilt a worldline more than 45 degrees. *This* is the beautiful visual payoff that we collect from the work of using "kilometers of time" in our spacetime diagrams.

The spacetime diagram in Figure 6 is not, of course, any kind of absolute truth. It is just the story as told in the reference frame of the farmer. Figure 7 shows the way the events would be plotted on a spacetime diagram using the reference frame of the

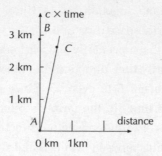

Figure 7. Spacetime diagram for the slower train's reference frame.

slower train. Not too surprisingly, in that reference frame, the train stays in the same place, the place called "0 kilometers" in that reference frame. Both events A and B occur at that place, but (of course) at different times. Event B occurs at 2.83 kilometers of time, or 0.00000943 second, after event A. This is slightly less than the 1/100,000 of a second that separated the events in the farmer's reference frame, but it is old news that the time between events is different in different reference frames. The coordinates of event C are also shown. These coordinates (like those of B) were computed using the Lorentz transformation, the "rotation in spacetime."

The key to understanding the mixing of time and space is contained in the comparison of Figures 6 and 7. In Figure 7 the line AB (that is, from A through B) is in a purely time direction; it has no space component. In Figure 6 AB is tilted; it does have a space component. By going to reference frames that are moving, one with respect to the other, we can tilt these upward directions in spacetime diagrams somewhat to the left and somewhat to the right, but only somewhat. There is a limit. If a line is primarily in the time direction (that is, more vertical than horizontal), we can never find a reference frame in which the line is primarily in a space direction (that is, more horizontal than vertical). A primarily time direction can be tilted, but it cannot be turned into a primarily space direction.

It also works out that a primarily space direction can never be turned into a primarily time direction. In Figure 6 take a look at the direction from event *B* to event *C*. Since these two events happen at the same time (in the farmer's reference frame), the direction from *B* to *C* is a purely spatial direction. It is horizontal. In Figure 7 the line between events *B* and *C* is no longer horizontal. There is now a time difference between the events, but the line is still more horizontal than vertical. If a direction in one spacetime diagram is primarily spatial, then in all spacetime diagrams it is primarily spatial.

The simple way of visualizing these things is to remember that the 45-degree lines in the spacetime diagrams are absolute barriers to rotation of space and time.

Time Machines

We all travel in time. With no effort and probably no choice, we move forward. The phrase "time travel," however, is associated with the idea of deviating from the beaten path and going backward in time. By exploiting the vocabulary we have just developed, we can describe this in a more useful way. Suppose you are present at some event *E*. Can you come back to the same location in space at a slightly earlier time? This possibility sounds wonderful, of course. You could go back and put tape over your mouth before making that stupid remark. You could stop yourself from investing in that high-technology company that seemed so promising a year ago. We won't dwell here on the issues of logical consistency that such fantasies raise; these will be the central focus of the essay by Igor Novikov. Here our concern will be narrower: to establish only the basic ideas along with the vocabulary. What is described here in the Introduction is only one mechanism for time travel—probably the simplest mechanism, or at least the simplest to describe. Novikov will describe a related, but different, mechanism that uses a strong gravitational field. Stephen Hawking will mention yet another mechanism involving cosmic strings.

An underlying feature of all mechanisms can be identified at the outset. If you could come back to the location of E at a time a few moments before event E, then by waiting a few moments you would be back at E. You would have visited the same event twice. It would be as if you started at some point on the equator and headed due east only to return to your starting point; your path would be closed. Time traveling from event E back to E would be a similar path in spacetime, and spacetime physicists call it a "closed timelike path or curve." Why "timelike"? Because as you move along this path, you are always moving forward in time. Your watch is always showing increasing numbers if it is digital and the hands are moving in the clockwise direction if it is analog. In your body the small number of radioactive nuclei are decaying; radioactive fragments are not clumping together in antidecay. Your heart is pumping in its usual way, not "backward in time" with blood flowing in the wrong direction. You are, alas, growing older, not younger. And it is your older self who returns (if that is the correct word) to event E in spacetime.

What does it take to travel on a closed timelike curve? The idea underlying a basic mechanism is contained in Figure 6. For simplicity, that diagram is repeated here (as Figure 8) but with only events B and C illustrated. Events B and C are at the same time (in the reference frame of the farmer who constructed this spacetime diagram), but event C is farther to the right than event B by $1/2$ kilometer. Now suppose, just suppose, that there is a secret tunnel, a shortcut in space for getting from the location of

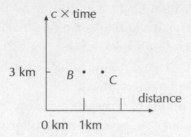

Figure 8. Two events that occur at the same time in a reference frame.

event B to that of event C. Suppose further that alongside the track, at the 1-kilometer mark (the location of event B), there is what looks like a well. And suppose that when you jump down that well, you find yourself instantaneously popping out at the 1.5-kilometer mark along the track (the location of event C).

We can get some feeling for what such a shortcut means. Likely, it is easier to get a grip on this shortcut than on the geometry of spacetime we will soon encounter. Here at least we are dealing only with a shortcut in space, not in spacetime. We *can*, more or less, picture space.

Let us consider a flat piece of paper. We probably want it to be infinite in extent, but we can only imagine and illustrate a finite portion of it. On the piece of paper, shown on the left in Figure 9, two dark spots, B and C, have been marked. The distance (that is, the shortest distance) between these two spots is, say, 1 m. But suppose that the paper is folded as shown on the right, and that there is a tiny tunnel, a kind of bridge or wormhole, from spot B to C. Aside from the possibility of adding this wormhole, nothing important about the paper changes when we fold it. In particular, all the distances measured along the paper (for instance, the length of any pencil line that we have drawn) remain the same when the paper is folded, so the geometry of the paper does not change. But the wormhole gives us a shortcut from B to C, and

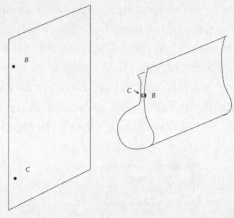

Figure 9. A wormhole shortcut between two points in a planar geometry.

we can make this wormhole shortcut as short as we want. This is the kind of wormhole that we want between the 1-kilometer and 1.5-kilometer marks alongside the train tracks. The model of the folded paper is not perfect, of course. It is a two-dimensional surface, and we are interested in a wormhole between two points in the three-dimensional spatial world of the farmer and the tracks. (We use only a single spatial dimension in our spacetime diagrams, but the wormhole between the two points along the track will need to exist in three dimensions.)

Until recently, spacetime scientists did not think all that much about wormholes connecting places in space. In the mid-1980s Kip Thorne became interested in the way spatial wormholes could be used to construct time machines, and soon many spacetime physicists were seen waving their hands and arguing about the mouths of wormholes. For an account of how this all started with a science fiction story by Carl Sagan, see Chapter 14 of Kip's popular book *Black Holes and Time Warps*.

But how do these spatial wormholes lead to time travel (or—to use the more dignified term—to closed timelike curves)? To see this, suppose some spacetime explorer witnesses event B and, by zipping through the spatial wormhole between B and C, finds herself at event C. At event C it turns out that a train, traveling at 100,000 km/s, happens to be going by. It is interesting to see what events B and C look like in the reference frame of that train. We have already drawn that spacetime diagram, as Figure 7. The relevant part of that figure is repeated in Figure 10.

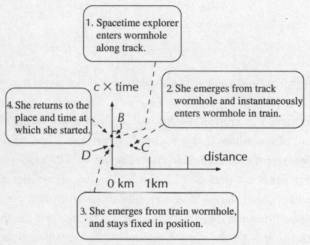

Figure 10. The spacetime for the reference frame of the moving train.

So far our spacetime explorer has traveled from B to C. We must now go on to suppose that the train is quite long and *also* carries a spatial wormhole. We are incredibly lucky: one end (or "mouth") of that train wormhole just happens to be right at the 1.5-kilometer mark as the explorer pops out of the track wormhole (event C). Wasting no time, she grabs hold of the train and jumps into the mouth of the train-based wormhole right in front of her; all this is activity at event C. Yet more incredible luck: the other mouth just happens to be at the 0-kilometer mark on the train. The explorer emerges from that mouth, having spent negligible time getting from one mouth of the very short wormhole to the other. In Figure 10 the event of her emergence from this mouth is indicated as event D. Note that events C and D are at the same vertical position in Figure 10. This is the graphical realization of the fact that negligible time passes between entering and emerging from the train wormhole, the entrance and emergence are therefore at essentially the same time *(in the reference frame of the train)*, and Figure 10 is, after all, a depiction of events in that reference frame.

She has now traveled on the path $B \rightarrow C \rightarrow D$. At around this time, as one might well imagine, her head is spinning, so she firmly plants her heels and stays put on the train. She is staying at the same location in the reference frame of the train, but not at the same time. By not moving in space, she "moves" in Figure 10 from event D to event B. She has completed a closed timelike curve and has gone back to B, the place and time she started.

All that is needed for this "round the spacetime" trip are two things: First, it is necessary to rotate directions in spacetime a bit. This is an immediate consequence of the Lorentz transformation and is questioned by very few physicists. Second, spatial wormholes are needed, and these have been questioned by many physicists. The question has not been answered yet, but it appears that the laws of physics do not allow spatial wormholes, and more generally do not allow time machines. The way quantum mechanical effects would (probably) destroy any budding wormhole is described by Stephen Hawking in his essay.

Why Does Spacetime Have a Geometry?

Though there are clear differences of detail between spacetime diagrams, maps of events, and maps of points in a two-dimensional plane, there are some fascinating similarities. The two questions here are: (1) are they really the same kind of thing, and (2) what does "same kind of thing" mean?

The mathematics of rotation—the equations of the rotational transformation—are an expression of the fact that there is a *geometry* to that plane. There is an orderly relationship of distances that is inviolate, and any way of describing distances must be compatible with that underlying geometric reality. The rotational mathematics is simply an inescapable tip of the iceberg. The geometry is the iceberg, the massive solid reality.

What about the Lorentz transformation? Is there an iceberg below this tip too? Is the Lorentz transformation only a description of a relationship that is guaranteed by an underlying geometry of events? There is no answer to this, since there is no unambiguous meaning to the reality underlying mathematics. Suppose we were given the mathematics of rotation and told that it accurately describes the relationship of measurements made by different reference frames (the observers on the bridge). A metaphysician could, while keeping a straight face, claim that the existence of the geometry is simply a mental construct to help us remember the rotational mathematics. It is not necessary to think of the geometry as real.

Most physicists have little patience with such arguments. In the case of the geometry of the plane, it seems like a pointless game to "pretend" that the geometry is not real. But the real defense of the geometry is not quite of the "I know what I see" variety. It's rather that the idea of there being a geometry is just so very useful. Not only does it help us remember the mathematics of rotation, but it also helps us to manipulate the mathematics and spot new relationships. If the geometry is not real, then it is so useful that its very usefulness makes it real.

When Einstein first proposed that the Lorentz transformation describes the relationship of event coordinates in different reference frames, he did *not* refer to any geometry. In his initial 1905 paper setting forth relativity, Einstein presented the Lorentz transformation as the only reality. It was Hermann Minkowski who pointed out to Einstein that these transformations could be viewed as the expressions of an underlying geometry, something we would now call the "Minkowski geometry of the spacetime of events." Minkowski's geometry was based on a way that he assigned a new kind of distance separating events, a distance that combined time and space. In different reference frames there will be disagreement about the time that separates the events and about the spatial distance between them, but there will be agreement about the Minkowski distance.

At first the Minkowski geometry seemed like an interesting construct, but quickly this construct became so useful that the idea that it was "only a construct" faded. Today Einsteinian relativity is universally viewed as a description of a spacetime of events with the Minkowski spacetime geometry, and the Lorentz transformation is a sort of rotation in that spacetime geometry.

Why Is the Geometry of Spacetime "Curved"?

One reason that Minkowski's introduction of the idea of spacetime geometry was so important is that it allowed Einstein to use the idea of curved spacetime geometry for describing gravity. The very phrase "curved spacetime" has such mystical imagery that it is too often shunned as incomprehensible. In at least one sense, however, the argument that gravity curves spacetime is not only comprehensible, it is compelling. What *does* have to be given up is any hope of visualizing curved spacetime with anything like the clarity of visualizing curved two-dimensional spatial surfaces. Do not consider spacetime theorists to constitute some priesthood of those who can actually *picture* curved four-dimensional spacetime. We can't. (I hope I'm not speaking only for myself here.) After all, it's space-*time!* And it's four-dimensional! We will draw diagrams, but they will be suggestive, often metaphoric, and sometimes poten-tially misleading. Our inability to picture curved spacetime slows, but does not limit, our ability to understand it. We still have mathematics, and we still have words.

The ideas start with the consideration of worldlines, the lines showing the events of an object as it moves forward in time. The worldlines in Figures 6 and 7 are the worldlines of trains in two different reference frames. These worldlines have a constant tilt (the angle at which they deviate from vertical). This means that the amount of distance they change per amount of time is always the same: they are constant-speed worldlines. Objects will not go at constant speed if forces are acting on them. Suppose that in the region of spacetime illustrated in Figure 11

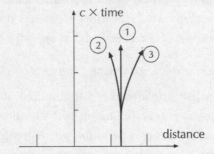

Figure 11. Particle worldlines in a region of spacetime
with electrical influence.

there is some strong electrical influence. For definiteness, let's say that it's caused by a large amount of positive electric charge somewhere off to the right of the figure.

An electrically charged object in the region of Figure 11 would accelerate (that is, it would change speed) due to the electrical influence. This change in speed, this acceleration, shows up in

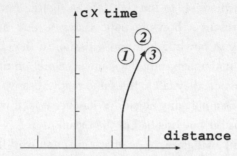

Figure 12. Particle worldlines in a region of spacetime
with gravitational influence.

the spacetime diagram as the changing tilt of a worldline. Since
worldline 1 in the figure is straight, it must tell the story of an
uncharged and therefore unaccelerating object. (In addition to
being straight, the worldline is vertical, which means that the
object is not only unaccelerated, it is also staying at one single
location in this reference frame.) The shape of worldline 2 tells
us that the particle it represents must be positively charged
since it is accelerating away from the positive charge (hidden
somewhere off to the right) creating the electrical influence.
Similarly, worldline 3 must illustrate the events of a negatively
charged particle. Looking closer we can see that worldline 3 is
more dramatically bent than worldline 2; its particle is experiencing a greater acceleration. Worldlines 2 and 3 might represent a proton and an electron. They have equal-sized charges of
opposite sign and the electron's much smaller mass would
account for the more dramatic bending of worldline 3.

A crucial point illustrated by Figure 11 is that each worldline
tells us something about the physical properties of the particle it
represents. Compare this now with worldlines representing
gravitational influence. Suppose the region of spacetime in
Figure 12 has a gravitational influence due to a large amount of

mass somewhere off to the right of the figure. Worldlines 1, 2, and 3 represent a bowling ball, a tissue, and an absanyon, respectively. A bowling ball and a tissue, in the absence of air resistance, undergo exactly the same acceleration under gravitational influence; they fall at the same rate. "Absanyon" I use here to mean "absolutely any object." Whatever it is, it will fall at the same rate as the bowling ball or the tissue.

The point of Figure 12 is that the bent worldlines tell us everything about the influence of gravity in this region of spacetime, and the same worldline describes the influence of gravity for any object. Einstein's very reasonable viewpoint was that the shape of the worldline by itself—not some "force"—should be the proper description of gravity. In Einstein's picture, objects experiencing only gravitational influence move only on special worldlines. The details of those worldlines contain the details of the gravitational influence.

What are those special lines in spacetime? In a gravity-free region of spacetime—in Minkowski spacetime—objects with no other influences move always in a fixed direction at a constant speed. Their worldlines are straight. So we know one example of special lines, and this gives us a clue of how to guess what we want in general. It turns out that straight lines don't exist in just any geometry. If we try to construct curves with all the properties of straight lines, we will usually fail. Consider the usual example (and a good one): the surface of a perfectly spherical Earth. Can we draw two lines that everywhere have a constant distance separating them the way parallel straight lines do? If we find that it *is* possible to draw a straight line in any direction through any point, we say that we are working in a "flat" space (or spacetime). Anything else, by definition, is curved.

In a curved space or spacetime, there is a simple generalization of the concept of a straight line: it is just the straightest possible curve that can be drawn. Such a curve has the fancy name "geodesic." When we look at a very small portion of a curved geometry, it looks almost flat. If a geodesic is drawn through that small portion, the geodesic will be almost straight.

For gravity to have its familiar properties, the worldlines of gravity-influenced objects cannot be truly straight lines. Consider a simple example: two satellites in orbit around Earth have a near collision, and a few orbits later have another near collision. This means that the worldlines of the satellites touch (or nearly touch) in two places. Straight lines cannot do this. The conclusion is inevitable: if it is to exhibit gravitational effects, then spacetime must be curved.

Although the mathematics of the special curves, the geodesics, is not trivial, neither is it terribly hard. Once a geometry is specified—that is, once a formula is given that tells the distances separating points in a space or spacetime—it is relatively easy to find the geodesics. In most (usually graduate) courses about Einstein's theory, the mathematics of the special curves comes early. What comes much later is the difficult part of the theory: the way the content of spacetime (the stars, the planets, and so forth) determines the spacetime geometry. Very fortunately we can skip that part without losing too much of the sense of the theory. We need only note that there is a mathematical prescription for determining the geometry of spacetime.

Gravitational Waves

Without knowing the detailed way in which matter curves spacetime's geometry, we know some features that the matter-curvature connection must have. A varying matter distribution must produce a varying curvature. Figure 13 is a suggestive illustration of the worldlines of a pair of binary stars, two massive stars in a tight orbit around each other. Note that as time moves

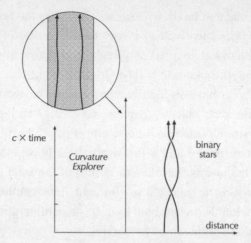

Figure 13. Gravitational waves from binary stars.

forward (that is, as we follow the worldlines upward in the diagram), the positions of the stars change during their mutual orbiting, and hence the "source" of gravitation is varying. This means that there will be variations in the spacetime geometry near the binary pair. Suppose that a scientific spacecraft, the fictitious *Curvature Explorer*, happens to be in the same corner of the galaxy as the binary pair and hence is in the region of varying spacetime curvature. What does this *mean?* What sign will the "curvonauts" in the spacecraft have of these variations?

The first answer that comes to mind is wrong. The curvonauts will not feel the spacecraft shake back and forth as it wends its way through the ruts of the curved spacetime. The reason why not lies in the question "shake back and forth with respect to what?" Against what "fixed" reference system would the motion of the spacecraft be wavering? If the spacecraft is not firing its correction jets, and if it is not being pummeled by micrometeorites, then the only influence on it is gravity. It is going along the straightest possible line (a geodesic). It is, in a sense, just riding spacetime the way a hot-air balloon rides the air. The balloon passengers can see themselves moving past the ground, but for the curvonauts there is nothing like fixed ground, just the curved spacetime.

The reason the curvonauts cannot perceive the spacecraft wavering suggests what they *can* see. They can perceive the motion of something only in comparison with something else. Let them focus very careful attention on two small objects inside (or outside) the spacecraft. With obsessive care let the curvonauts shield the objects from all influences except the all-pervasive, unshieldable influence of gravity. The curvonauts will then be able, in principle, to detect a wavering distance between the objects. What they are observing is the spacetime equivalent of the wiggles in the distance between two straightest possible lines drawn on a rippling surface.

In the details of the theory, there are many similarities between the effects due to the binary stars and the effects due to oscillating electric charges. The charges create varying electrical and magnetic influence that have certain "wave" properties. In particular, the strength of the oscillation-caused electromagnetic waves decreases in a very particular and simple way with increasing distance from the oscillating charges. When you move twice as far from the charges, the oscillating influence you measure is half as large. A second important property is that the oscillating influences propagate at speed c. The spacetime oscillations due to the binary stars, or to any varying gravity source, also have these two properties, and are called gravitational waves.

Horizons and Black Holes

Presently we are going to consider very strong gravitational fields and very strongly curved regions of spacetime, but things will be clearest if we start with no gravity and with Minkowski

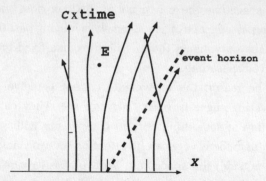

Figure 14. A simple event horizon in Minkowski an spacetime

spacetime. Figure 14 is a spacetime diagram for Minkowski spacetime, drawn with the usual convention so that a 45-degree worldline would be moving at the speed c. The many worldlines shown are not meant to depict objects moving only under gravitational influence. Since the special curves of Minkowski spacetime are straight, the worldlines in the figure must represent objects with nongravitational forces acting. We could suppose, for instance, that the nongravitational forces causing the worldlines to curve are the thrust of rocket engines and that the worldlines are those of the rockets.

Figure 14 contains another interesting feature: a dashed 45-degree line. We can think of this line as representing a wall that is infinite in the y and z directions (not shown in the figure) and is moving at speed c in the x direction. What is interesting about the dashed line is that worldlines can only cross it from right to left, never from left to right. This can be seen in the fact that worldlines can never deviate from the vertical by more than 45 degrees. Equally well we can understand this as necessary because the "wall" represented by the dashed line is moving from left to right at speed c. To cross the wall from left to right, an object would have to be moving faster than c. So the dashed line is a one-way barrier that divides spacetime into two regions. Any object on the left side of the dashed line can never get to the right side.

There is a property of the dashed line that is even stranger: an object on the right side of it can never have any knowledge of any event on the left. Consider event E on the left side. The worldline of any signal sent out from event E, whether by postcard or light signal, can never cross the dashed line to be received by some observer on the right. For this reason, the dashed line is called an event horizon. Just as sailors cannot see ships over the horizon, observers on the right of the horizon cannot "see" (get information from) events on the left.

A somewhat differently shaped event horizon is shown in Figure 15. The dashed circles represent three different times of a

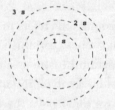

Figure 15. An expanding spherical horizon in Minkowski spacetime.

spherical surface that increases its radius by 300,000 kilometers every second. The spherical surface is therefore moving outward at speed c. Clearly, any object inside the expanding sphere cannot cross the sphere; clearly, observers outside the expanding sphere can receive no information about events occurring inside the sphere. The expanding sphere therefore divides spacetime into two regions, just as the dashed line does in Figure 14.

We have seen two examples of event horizons, but they were mathematical constructs and showed us nothing about gravity. We now consider a very strong, spherically symmetric gravitational field, in which gravity is pulling toward some central point with equal strength in all directions and is unchanging in time. The corresponding spacetime is called "Schwarzschild spacetime" in Einstein's theory.[5] Figure 16 shows a new kind of

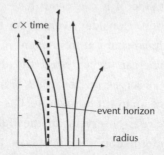

Figure 16. A Schwarzschild event horizon.

spacetime diagram. Here the vertical axis, as usual, indicates time, but now the horizontal axis refers to radius.

There is a big difference between this spacetime diagram and the usual one. This diagram illustrates a spacetime region with strong gravitational fields, so it represents a spacetime that is curved. In this region we don't have the clear meaning of time and distance that we had in Minkowski spacetime. This is a difficult but important concept. The symbol "$c \times$ time" is a convenient coordinate for marking spacetime events, but it is not the time that a clock ticks.[6] Similarly "radius" is not a measured distance but a convenient coordinate. It is worth mentioning why it is convenient: it gives the correct area with the common formula for surface area. That means that a spherical surface at (coordinate!) r has the area $4\pi r^2$, the usual relationship we are supposed to learn in high school. This formula is not a trivial statement since the spacetime is curved. The same coincidence with simplicity does not apply to radial distance; the radial coordinate r that gives the right area does *not* correctly measure radial lengths.

Figure 16, then, is representing a region of spacetime in convenient coordinates. We have lost the feature of earlier spacetime diagrams that 45-degree lines represent motion at speed c. The worldlines in this figure are rocket worldlines; they show spacetime paths of objects not limited to feeling only gravitational influence. There is a general tendency for the worldlines to bend to the left (a smaller radius), due to the pull of gravity toward the center. There are worldlines that show some bending to the right; these represent objects whose rocket engines exert more force on the object than gravity does. All of this is expected. What is not expected is the dashed vertical line. It is an event horizon. No worldline can cross from left to right. No event to the left can transmit information to any receiver on the right. The dashed line represents a spherical surface that has the same horizon properties as the expanding spherical surface of Figure 15, except that now the spherical surface is not expanding. It is, after all, at an unchanging value of radial coordinate, so it has an unchanging surface area. What else could "not expanding" mean?

The region inside the horizon, a region that cannot be "seen" by any observer on the outside of the horizon is called, with obvious justification, a black hole. This spherical black hole, in particular, is called a "Schwarzschild black hole" and its boundary a "Schwarzschild event horizon." It is the mathematically simplest solution of Einstein's theory that has black hole properties. Another relatively simple black hole solution is the "Kerr black hole." It has a rotational axis and sense of rotation that a spherical Schwarzschild black hole does not. For a Kerr black hole, like a Schwarzschild black hole, the area of the event horizon is unchanging. Black holes do not have to be unchanging in time. The event horizons that define them can change shape and can grow. But there must be a limit to the growth. If the horizon could grow so that there would eventually be no "outside," as in the case of Figure 15, then the inside would not be a black hole. A black hole needs to have an exterior forever, an exterior that never can know about what is in it.

Bon Voyage

You are about to sail out into interesting seas of ideas, and you are now packed—though rather lightly—for that journey. This introduction has been a succinct traveler's guide for what is to come; it has given you a sketchy map of the new realms, along with a phrase book with the basic vocabulary of those who work in those realms. Like any brief guide, it is not where the excitement is to be found. For that you have to make the journey on the pages to follow.

Notes

1. Kip S. Thorne, *Black Holes and Time Warps: Einstein's Outrageous Legacy* (W. W. Norton, New York, 1994).
2. Edwin F. Taylor and John Archibald Wheeler, *Spacetime Physics* (W. H. Freeman, San Francisco, 1992).
3. Nick Huggett, *Space from Zeno to Einstein: Classic Readings with a Contemporary Commentary* (MIT Press, Cambridge, Mass., 1999).
4. The speed c is usually called the "speed of light," but this can be misleading so I avoid it. Calling c the speed of light is too often interpreted to mean that light propagation is somehow responsible for relativistic effects, when in fact the propagation of light signals has nothing to do with these effects. Light does (in a vacuum) move at the speed c, but that is a consequence of the role c plays in the structure of spacetime. If there were no electromagnetism, there would still be c.
5. The Schwarzschild spacetime is named after Karl Schwarzschild, who in 1916 showed that it is a solution of Einstein's equations. The radial and time coordinates described here are often called the "Schwarzschild coordinates."

6. The traditionally used time coordinate is convenient because, with this meaning of time, the gravitational field is not changing in time. This is intuitively disturbing because our intuition tells us that the gravitational field either *is* or *is not* changing in time. This fallacious intuition is based on the idea—hard to shake—that time is absolute, and could be measured by a universal clock ticking the same for everyone and everything.

CAN WE CHANGE THE PAST?

Igor Novikov

In this essay I will explore various aspects of time machines, despite the fact that the essay by Stephen Hawking in this volume explains that time machines, in all likelihood, are physically impossible. There are two reasons for my ignoring Hawking's prediction. First, in 1895, another outstanding physicist, Lord Kelvin, then president of the Royal Society, claimed that "heavier-than-air flying machines are impossible." Lord Kelvin's claim was based on the best understanding of physics at that time. However, as we know, the first flight by the Wright brothers was achieved in 1903, only a few years later. In a similar way, our present understanding of time machines may be incomplete. The second reason is one that Kip Thorne has pointed out many times: even if the laws of physics forbid time machines, the effort to understand them may teach us much by helping us to sharpen our understanding of causality.[1]

Let us therefore assume that time machines are possible in principle, and explore the consequences. First, time machines may be dangerous. Indeed, if someone could travel from our time back

neutron star

into the past, then that person could probably change the past. If so, as a result, he would change all of subsequent history. For example, a person who travels back in time to the beginning of the universe could change the physical conditions at that period, and as a result change the whole history of the universe. The explosion of a hydrogen bomb is nothing in comparison to such a possibility.

Is it in fact possible to use a time machine to change the past? We can imagine time as a river that flows from the past into the future, never changing direction, carrying in its flow all events. For many years people believed time could not be slowed down or accelerated. However, at the beginning of the twentieth century, Albert Einstein discovered that time is not immutable. Strong gravitational fields—for example, the strong gravitational field of a neutron star—slow down the pace of time. Clocks in the strong gravitational field near a neutron star's surface tick more slowly than do clocks far away. Observers at a distance from a neutron star can in principle see the clocks' slowdown.

According to general relativity, the modern theory of gravity, spacetime should be warped in strong gravitational fields. What this means is illustrated in Figure 1a, which shows the space-

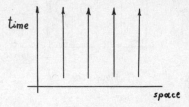

Figure 1a.

time continuum with space horizontal and time vertical. Strong gravitational fields give rise to indentations, or wells, in the surface, as illustrated in Figure 1b. The key idea underlying time

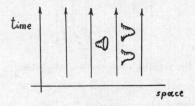

Figure 1b.

machines can now be explained. Imagine that the tops of two different indentations touch each other (Figure 1c) and join up

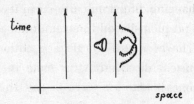

Figure 1c.

to form an arch or tunnel, as in Figure 1d. The effect of such an arch is that a piece of the river of time separates from the main

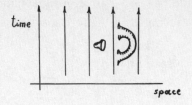

Figure 1d.

flow, goes through the arch, and rejoins the main flow at an earlier time than it entered the arch (Figure 1e). A human being on

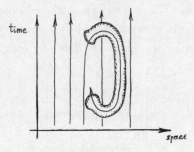

Figure 1e.

such a path would be carried along with the river of time through the tunnel, all the while becoming older, until he reappears at the main flow to the past of (that is, at an earlier time than) where he entered the arch, as in Figure 1f . Thus, he could meet an earlier version of himself. Stephen Hawking in this volume explains that such spacetime structures can arise as mathematical solutions of Einstein's field equations. While most scientists dismiss these solutions as mathematical curiosities, Kip Thorne, together with his younger colleagues, has recently seriously investigated such solutions.

This essay will discuss the following three issues related to time machines. First, how could time machines be created? Second, is it actually possible to change the past using a time

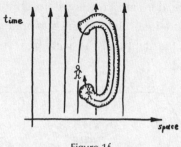

Figure 1f.

machine? Third, what happens to our notions of causality and free will?

The first issue is how can time machines be created. This issue requires discussion of curved or warped spaces. Since it is very difficult to imagine or visualize curved three-dimensional spaces, let's use instead as an analogy curved two-dimensional spaces, populated by the two-dimensional person of Figure 2a.

Figure 2a.

Consider a two-dimensional star in this space, whose gravitational field is rather weak. The star's gravitational field will appear as a shallow well, as illustrated in Figure 2b. Now sup-

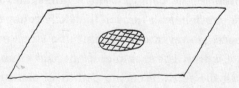

Figure 2b.

pose we squeeze the star, causing it to contract. During this process the gravitational field of the star increases in strength, and as a result the depth of the well increases, as shown in Figures 2c and 2d.

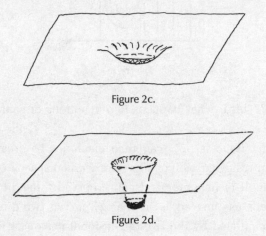

Figure 2c.

Figure 2d.

Now, let's suppose that we have two such wells and that the bottoms of these two wells touch each other. This creates a structure called a "wormhole," illustrated in Figure 3. It consists of two openings, or mouths, and a tunnel or arch between them. Let's assume that we can stabilize such a structure so that it's static (that is, unchanging) or almost static. Then, a two-dimensional being can go from one mouth to the other in two ways: through the "external" space or through the tunnel of the wormhole. As can be seen from Figure 3, the route through the wormhole is longer than the route through the external space. However, one can imagine situations in which the opposite is true. For example, if one had a tunnel through the center of the earth, then the path over the surface of the earth from one opening to the other would be longer than the path through the center along the tunnel. A similar situation

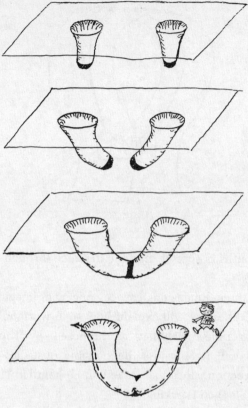

Figure 3.

can apply to our two-dimensional models of warped space. As illustrated in Figure 4, the two-dimensional space can be bent in such a way that the distance through the corridor between

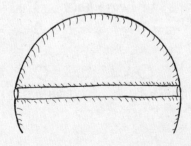

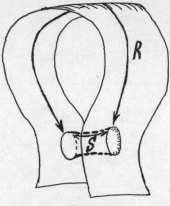

Figure 4.

the two mouths is shorter than the distance through the external space.

In real, three-dimensional space, one can imagine similar wormholes. Shortly I will explain how such wormholes can be turned into time machines, as proposed by Kip Thorne. However, I will not discuss the possible obstacles impeding their construction, since that issue will be handled later in this volume by Stephen Hawking.

Let's now imagine that in three-dimensional space we have a wormhole consisting of two openings, mouth A and mouth B, and a corridor between them. Of course, the corridor does not reside in normal three-dimensional space; rather, it can be

thought of as living in a higher-dimensional hyperspace. Let's assume that the distance along the corridor between the mouths is much shorter than the distance between the mouths in the external space. If there is a guy at mouth *A* and a girl at mouth *B*, and the guy looks at the girl through normal space, then he will perceive her to be at a large distance, many miles or even light-years away. However, if he looks at her through the wormhole, it's possible in principle for her to be only a few meters away.

Now the wormhole can act as a "space machine," because the guy can step through it to be next to the girl, but it is not yet a time machine.

Here is how such a wormhole *can* be converted into a time machine. Suppose we put two clocks near the two mouths,

clock A near mouth A and clock B near mouth B, and that the clocks are initially synchronized. Let's further place mouth B in the strong gravitational field of a neutron star. Recall that the pace of time depends on the strength of the gravitational field,

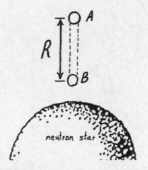

and that time flows slowly near the surface of a neutron star, so time flows slowly near mouth B. If mouth A is farther from the neutron star than mouth B is by a distance R, then the difference between the pace of time at mouths A and B will be proportional to R. After some time has passed, the two clocks will have different readings. Clock *A* might read five minutes to twelve, whereas clock *B*, having ticked more slowly, might read twenty minutes to twelve, for instance.

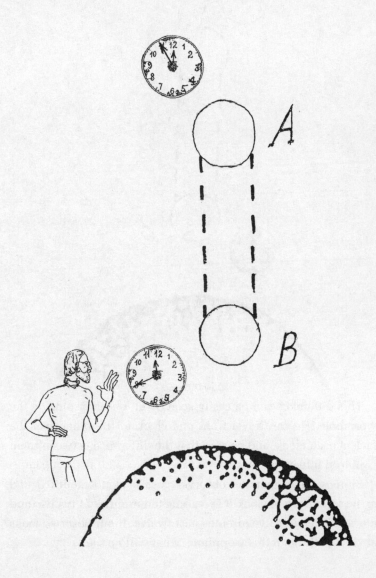

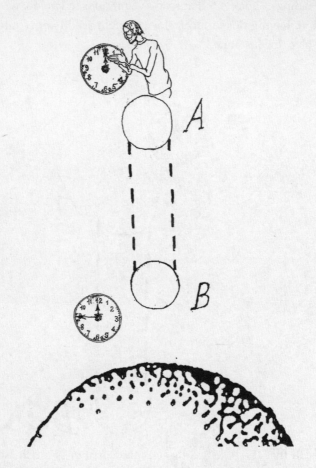

This difference can be easily seen by an observer outside the wormhole. He can travel from one clock to the other, feel the face of each clock, and be sure that the difference between them is indeed fifteen minutes.

Suppose now that the observer goes back to mouth B and arrives there when clock B is reading ten minutes to twelve, and clock A is reading five minutes past twelve. If our observer looks at clock A through this wormhole, what will he see?

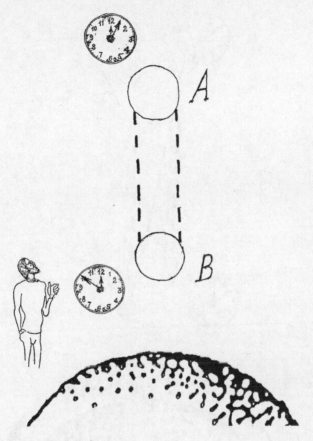

Recall that the distance through the corridor is much smaller than the distance between the same points through outer space. The observer therefore sees the second clock to be nearby, so both clocks are practically side by side, the distance between being only a few meters. Also recall that the difference in the ticking rate of the two clocks is proportional to the distance between them. Through the wormhole, that distance is practically zero, and so there is no difference in the ticking rate of the two clocks. They will tick in unison and they will show identi-

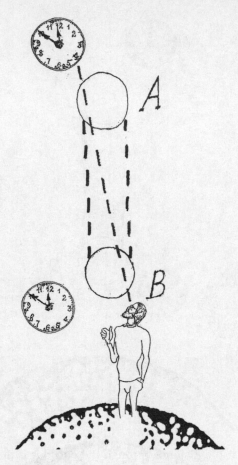

cal time always, as perceived by observers looking through the wormhole. Therefore, if the observer looks through the wormhole, he sees clock A showing the same time as clock B, since during the entire experiment the two clocks have ticked in unison. He will see both clocks showing ten minutes to twelve. However, this means that in looking through the corridor, the observer is looking into the past, since if he looks at clock A outside the wormhole he sees it reading five minutes past twelve.

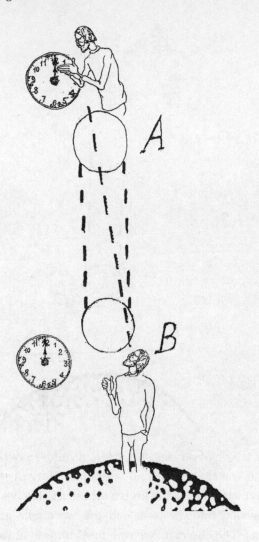

Through the wormhole, he sees ten minutes to twelve, so he sees the past. In fact, if he waits for ten minutes, he will see himself at clock *A*, since he came to clock *A* and touched it while it was reading twelve o'clock.

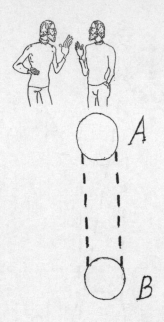

Now the observer can start at ten minutes to twelve at clock B and can travel through the wormhole. He will then arrive at mouth A when clock A reads ten minutes to twelve. He therefore has traveled into the past, and the wormhole acts as a time machine. The observer can even meet himself at twelve o'clock, at mouth A.

Two remarks should be made here. First, this time machine can in principle be very powerful. The longer we wait with mouth B in the strong gravitational field, the greater the difference in time between clocks A and B becomes. We can arrange for the time machine to transport us many hours or even years

into the past. Second, if we remove the wormhole mouths from the vicinity of the neutron star, away from the strong gravitational field, they will continue to work as a time machine. Therefore, we can in principle construct a time machine that consists of two mouths, A and B, and a very short corridor in some additional dimension joining them, such that an observer entering mouth B will appear from mouth A in the past. As explained in Stephen Hawking's essay, this means that he can meet a younger version of himself in the past.

It's clear that time machines like this can give rise to paradoxes. Suppose I enter mouth B and reappear at mouth A, in the past, before I enter mouth B. Then, there will be two versions of myself, a younger one about to enter mouth B, and an older one who has

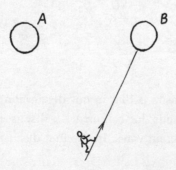

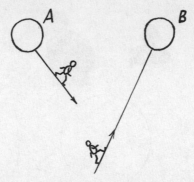

just come out of mouth *A*. Suppose I take a knife and kill the younger version of myself. Then it becomes impossible for the younger version of myself to continue on my way to mouth *B* and emerge from mouth *A* to perform the homicide. This is a paradox. Alternatively, I can use a more powerful time machine, travel back into the distant past, and kill my grandmother before the birth of my mother, which leads to a similar paradox.

Does this paradox demonstrate that time travel is impossible?

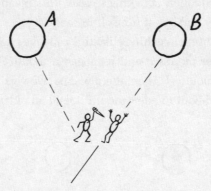

Not at all. The reason is that in our discussion of the paradox I committed a serious logical error. I discussed the situation twice, in two different ways. In the first discussion, I discussed my journey to mouth B assuming there was no meeting with the older version of myself from the future. In the second discussion, I discussed the same journey but assumed that the first discussion was correct and that therefore I could travel back in time and that therefore there was a meeting. The error is the assumption in the first discussion that there was no meeting. If the meeting happened, it happened. So we should take into account the consequences of the meeting from the very beginning. Therefore, even if I am not killed, when I travel into mouth B, then I will remember the meeting with my younger self when I come out of mouth A.

So what actually happens in this scenario, when we analyze it without any incorrect assumptions? Here we run into a complication: physicists cannot calculate explicitly the actions of human beings, because human beings are too complicated. It's an issue for psychology probably, or zoology, but not for physics. However, physicists can model similar, apparently paradoxical situations involving simple physical bodies. Using such models, physicists can calculate what happens to the bodies and determine how the paradox is resolved.

So, let's try to model this paradox using simple physical objects. Imagine a billiard ball rolling on a table with an opening or pocket. It's not difficult for somebody to the push the ball in the direction of the opening, so that the ball rolls on the table and eventually falls down into the opening. Suppose there is a second billiard ball traveling along a path that intersects the path of the first ball before the first ball reaches the opening. Then there will be a collision. If the collision is strong enough, the change in direction of the first billiard ball will be large

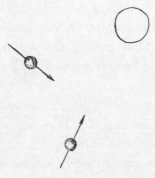

enough that after the collision the first ball will move on a completely different trajectory and miss the opening entirely.

Suppose we have a time machine with mouths A and B as described earlier, so that if someone enters mouth B, he will reappear at mouth A in the past. We take just one billiard ball and send it on a trajectory toward mouth B. The billiard ball is traveling through empty space now rather than on the surface of a billiard table, but this difference is unimportant. The ball will move through space toward mouth B. However, before it reaches mouth B, it will appear from mouth A because of the time machine effect. So, there will be two versions, or incarnations, of the same billiard ball: a younger version and an older version. We can arrange the initial push of the ball in such a way that the paths of the two versions intersect and so that the two versions come to the meeting point almost simultaneously. What will be the result? Just as with the billiard balls on the billiard table, we

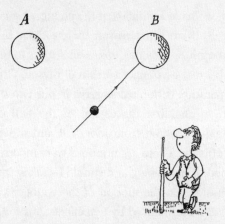

can arrange the initial push such that the older billiard ball hits
the younger one strongly enough to deflect its trajectory so that
the younger one never reaches mouth B.

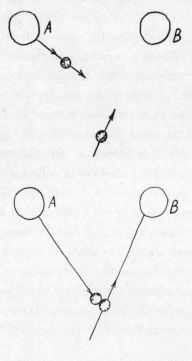

Once again, we have a paradox: if the younger billiard ball never enters mouth *B*, there is no reason for the older billiard ball to appear from mouth *A*. This paradox is analogous to the earlier paradox in which I met a younger version of myself. Once again, the origin of the paradox is the logical error in our two discussions of the situation. In the first discussion of the ball's motion, we assumed no ball appeared from mouth *A* and did not take into account the collision. These assumptions were incorrect: if the collision happened, it happened, and should be taken into account in the first discussion. So the motion of the younger version of the billiard ball will be influenced by two effects: our initial push, and the effect of the collision with the older billiard ball.

So what is the resolution of the paradox? What actually happens if we send the billiard ball toward mouth *B* with an appropriate initial push? In this situation, we can calculate what happens because billiard balls are a very simple mechanical system. If we take into account the collision from the very beginning, then the collision is very weak, just a slight touch between the two balls that nudges the younger ball only slightly. The younger ball then moves along a trajectory slightly different from our expectation, but still enters mouth *B*. It reappears from mouth *A* in the past and continues along its motion, still on a trajectory that differs only slightly from the trajectory it would have traveled on had it not suffered a collision. The result of the slight difference in trajectory is that the collision with the younger version of itself is not a strong collision, but rather a weak collision, a glancing blow. Therefore we have a consistent solution. Although we tried to arrange a strong collision, we find that in fact the collision is weak, if we analyze the situation correctly taking into account the collision from the very beginning. This consistent solution can be obtained from a rigorous mathematical calculation, and was first discovered by Kip Thorne.

We see now that there are no contradictions or paradoxes, and moreover, there are not two different versions of the collision event. There was only one collision, only one history of events. If something happened, it happened. Events can be influenced by other events in the future, as well as other events in the past, so the flow of events can be complex. However, there is only one flow of events, so the past cannot be changed once it has occurred.

One might object that we have analyzed a trivial mechanical system, and that other more complex systems lead to drastic paradoxes that cannot so easily be resolved. For example, let's suppose that instead of a billiard ball we have a bomb whose surface is covered with fuses, so that even the slightest touch of

a ball with a bomb and fuses

the bomb's surface would cause it to detonate. At first sight, it looks as if there is no self-consistent flow of events in this situation. Indeed, suppose we push the bomb in an appropriate way in the direction of the wormhole's mouth B. The bomb should appear from mouth A in the past and move along a trajectory to the meeting point. If it meets an earlier version of itself, even a glancing collision will cause an explosion. The younger version of the bomb will be destroyed, making it impossible for the

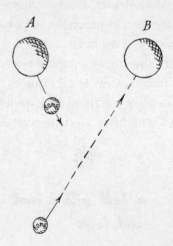

younger version of the bomb to continue its motion toward
mouth B, and for the older version to reappear from mouth A. So
there's a paradox.

The resolution of this paradox is the following. While the bomb
moves along its trajectory toward mouth B, something emerges
from mouth A—not an older version of the bomb, but a fragment

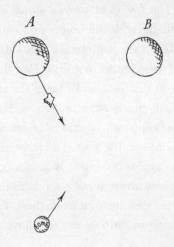

of the bomb. (It will become clear shortly why a fragment appears from mouth *A*.) The fragment moves along a trajectory to the meeting point, collides with the bomb, and causes an explosion. Fragments of the bomb are scattered in many different directions, and at least one of them enters mouth *B*. That fragment then reappears at mouth *A* in the past, and causes the original explosion. Again we have a self-consistent situation without any contradictions or paradoxes. However, we can also see very explicitly the influence of the future on the past. The fragment from the future is the cause of the explosion, but it is also a consequence of the explosion. This is very unusual but not contradictory.

We can draw two conclusions from these examples. First, in the presence of time machines, we can have very strange and unusual physical processes taking place, but no contradictions. Second, any event (like the explosion) cannot be changed. It happened only once, and it is impossible to have two histories, one in which the event occurs and another in which it did not occur.

Let's now reconsider the apparently paradoxical situations involving human beings. Would it be possible for you to use a time machine to travel to the past and kill a younger version of yourself? The answer is no, it's impossible. That would lead to a paradox, and we saw earlier that paradoxes do not occur. Therefore physical laws should prevent you from killing a

younger version of yourself. As Kip Thorne has said, if you try to kill a younger version of yourself, or your grandmother, something must stop your hand. Its not possible for physicists to calculate exactly what stops your hand, because human beings are very complex, although we can calculate what happens for simple physical objects as we've done here.

This means that our free will must be constrained. If I meet with a younger version of myself and wish to kill that younger version, then the laws of physics will prevent me from doing so.

"Miss! Oh, Miss! For God's sake, stop!"

Such a constraint on our free will is unusual and mysterious but not completely without parallel. For example, it can be my will to walk on the ceiling without the aid of any special equipment. The law of gravity prevents me from doing so; I will fall down if I try, so my free will is restricted. Of course, in the case of a time machine, the nature of the restriction on free will is different, but not essentially different.

In conclusion, the question of whether time machines can exist is still unresolved. However, even if time machines are forbidden by the laws of physics, it is still worthwhile to think about the issues they raise, because it can bring us new insights about the nature of time, causality, and other aspects of physics. Finally, we cannot change the past. We cannot send a time traveler back to the Garden of Eden to ask Eve not to pick the apple from the tree.

Note

1. Kip S. Thorne, "Closed Timelike Curves," in *General Relativity and Gravitation 1992: Proceedings of the 13th International Conference on General Relativity and Gravitation*, ed. R. J. Gleiser, C. N. Kozameh, and O. M. Moreschi (Institute of Physics Publishing, Bristol, England, 1993), pp. 295–315.

CHRONOLOGY PROTECTION: Making the World Safe for Historians

Stephen W. Hawking

This essay will be about time travel, which has become an interest of Kip Thorne's as he has become older. (Is this a coincidence?) But to speculate openly about time travel is tricky. If the press picked up that the government was funding research into time travel, there would either be an outcry at the waste of public money, or a demand that the research be classified for military purposes. After all, how could we protect ourselves if the Russians or Chinese had time travel and we didn't? They could bring back Comrades Stalin and Mao! So there are only a few of us who are foolhardy enough to work on a subject that is so politically incorrect, even in physics circles. We disguise what we are doing by using technical terms like "closed timelike curves," which is just code for time travel.

The first scientific description of time was given in 1689 by Sir Isaac Newton. Newton held the Lucasian Chair at Cambridge that I now occupy (though it wasn't electrically operated in his time). In Newton's theory, time was absolute and marched on relentlessly. There was no turning back and returning to an earlier age. The situation changed, however, when Einstein formulated his general theory of relativity in 1915. Time was now linked with space in a new entity called spacetime. Spacetime was not an absolute fixed background in which events took place. Instead, space and time were made dynamic by the Einstein equations, which described how they were curved and distorted by the matter and energy in the universe. Time still increased locally, but there was now the possibility that spacetime could be warped so much that one could move on a path that would bring one back before one set out. A few years ago, the BBC made a film with Kip and me showing what this kind of time travel might be like. They used trick photography to depict "wormholes," hypothetical tubes of spacetime that might connect different regions of space and time. The idea is that you step into one mouth of the wormhole and step out of the other mouth into a different place and at a different time.

Wormholes, if they exist, would be ideal for rapid space travel. You might go through a wormhole to the other side of the galaxy and be back in time for dinner. However, one can show that if wormholes exist, you could also use them to get back before you set out. One would then think that you could do something, like blowing up the rocket on its launch pad, to prevent your setting out in the first place.

This is a variation of the grandfather paradox: what happens to you if you go back in time and kill your grandfather before your father was conceived?

Of course, this is a paradox only if you believe you have free will to do what you like when you go back in time. I'm not going to get into a philosophical discussion of free will in this essay. Instead, I shall concentrate on whether the laws of physics allow spacetime to be so warped that a macroscopic body like a spaceship can return to its own past. According to Einstein's theory, a

time travel ◀━━━━━━━━▶ closed timelike curves

spaceship necessarily travels at less than the local speed of light, and it follows what is called a "timelike curve" through spacetime. Thus one can formulate the question in technical terms: does spacetime admit timelike curves that are "closed," that is, which return to their starting point again and again?

Does spacetime admit closed timelike curves?

1. Classical theory
2. Semi-classical theory
3. Full quantum gravity

There are three levels on which we can try to answer this question. The first is Einstein's general theory of relativity. This is what is called a "classical theory." That is to say, it assumes the universe has a well-defined history, without any uncertainty. For classical general relativity, we have a fairly complete picture, which I shall describe.

We know, however, that classical theory can't be quite right, because we observe that matter in the universe is subject to fluctuations, and its behavior can't be predicted precisely. In the 1920s, a new paradigm called "quantum theory" was developed to describe these fluctuations, to quantify the uncertainty. We can therefore ask the question about time travel on a second level, called the "semiclassical theory." In this, quantum matter fields are considered on a classical spacetime background. Here the picture is less complete, but at least we have some idea how to proceed.

Finally, there is the full quantum theory of gravity, whatever that may be. Here it is not clear even how to pose the question, Is time travel possible? Maybe the best one can do is to ask how observers at infinity would interpret their measurements. Would they think that time travel had taken place in the interior of the spacetime?

Let's start with the classical theory. Flat spacetime does not contain closed timelike curves, nor do other solutions of the Einstein equations that were known early on. It was therefore a great shock to Einstein when, in 1949, Kurt Gödel, better known in mathematics for Gödel's theorem, discovered a solution that represented a universe full of rotating matter, with closed time-

---

Gödel Universe

A spacetime containing
rotating matter with closed timelike
curves through every point

---

like curves through every point. Gödel's solution required a cosmological constant, which may or may not exist in nature, but other solutions were subsequently found without one.

A particularly interesting case is two cosmic strings moving at high speed past each other. As their name suggests, "cosmic strings" are objects with length but a tiny cross section. They are predicted to occur in some theories of elementary particles. The gravitational field of a single cosmic string is flat space with a wedge cut out, such that the sharp edge is at the string. Thus if one goes in a circle around a cosmic string, the distance in space

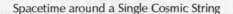

Spacetime around a Single Cosmic String

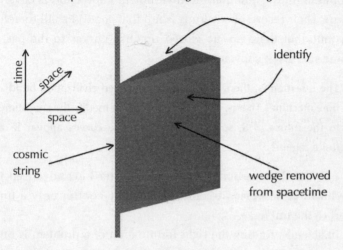

is less than one would expect, but time would not be affected. This means that the spacetime around a single cosmic string does not contain any closed timelike curves. However, if there is a second cosmic string that is moving with respect to the first, the wedges cut out for each will shorten both spatial distances and time intervals. If the cosmic strings are moving at nearly the speed of light relative to each other, the saving of time in going around both strings can be so great that one would arrive back before one set out. In other words, there are closed timelike curves that one can follow to travel into the past.

Cosmic string spacetime contains matter that has positive energy density and is physically reasonable. However, the warping that produces the closed timelike curves extends all the way out to infinity, and back to the infinite past. Thus these spacetimes were created with time travel in them. We have no reason to believe that our own universe was created in such a warped fashion, and we have no reliable evidence of visitors from the future. (I'm discounting the conspiracy theory, that UFOs are from the future, and that the government knows and is covering it up. Their record of coverups is not that good!) I shall therefore assume that there are no closed timelike curves to the past of some surface of constant time, S.

The question is, then, Could some advanced civilization build a time machine? That is, could that civilization modify the spacetime to the future of S, so that closed timelike curves appear in a finite region?

I say "a finite region" because, no matter how advanced the civilization becomes, it could presumably control only a finite part of the universe.

In science, finding the right formulation of a problem is often the key to solving it, and this is a good example. To define what is meant by a finite time machine, I went back to some early work of mine. I had defined the future Cauchy development of S to be the set of points of spacetime where events are determined completely by what happens on S. In other words, it is the region of spacetime where every possible path that moves at less than the speed of light comes from S.

However, if an advanced civilization manages to build a time machine, there will be a closed timelike curve C to the future of S. C will go around and around in the future of S, but it will not go back and intersect S. This means that points on C will not lie in the Cauchy development of S. Thus S will have a Cauchy horizon H, a surface that is a future boundary to the Cauchy development of S. I had introduced the concept of a Cauchy horizon about the time I first met Kip, way back in prehistory, shortly after the Ark!

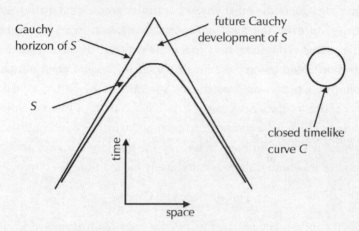

Future Cauchy Development of *S*

Cauchy horizon of *S*

future Cauchy development of *S*

S

time

space

closed timelike curve *C*

Cauchy horizons occur inside some black hole solutions, and in a solution that physicists call "anti–de Sitter space." However, in these cases, the light rays that form the Cauchy horizon start either at infinity or at singularities. To create such a Cauchy

horizon would require either being able to warp spacetime all the way out to infinity, or having a singularity in spacetime. Warping spacetime all the way to infinity would be beyond the powers of even the most advanced civilization, who would be able to warp spacetime only in a finite region. The advanced civilization could assemble enough matter to cause a gravitational collapse, which would produce a spacetime singularity, at least according to classical general relativity. But the Einstein equations could not be defined at the singularity, so one could not predict what would happen beyond the Cauchy horizon, and in particular, whether there would be any closed timelike curves.

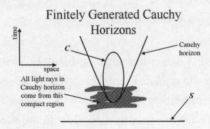

Finitely Generated Cauchy Horizons

I shall therefore take as my criterion for a time machine what I call a "finitely generated Cauchy horizon." This is a Cauchy horizon generated by light rays that all emerge from a compact region. In other words, they don't come in from infinity or from a singularity, but from a finite region containing closed timelike curves, the sort of region our advanced civilization is supposed to create.

Adopting this definition as the footprint of a time machine has the advantage that one can use the machinery of causal structure that Roger Penrose and I developed to study singularities and black holes. Even without using the Einstein equations, I can show that, in general, a finitely generated Cauchy horizon will contain a closed light ray, a light ray that keeps coming back to the same point over and over again.

Closed Light Ray

Talk about déjà vu! Moreover, each time the light comes around, it will be more and more blueshifted, so the images get bluer and bluer. The light rays may get defocused sufficiently each time around, so the energy of light doesn't build up and become infinite. However, the blueshift will mean that a particle of light will have only a finite history, as defined by its own measure of time, even though it goes around and around in a finite region and does not hit a curvature singularity. One might not care if a particle of light completed its history in a finite time. But I can also prove that there would be paths moving at less than the speed of light and having only finite duration. These could be the histories of observers, who would be trapped in a finite region before the Cauchy horizon, and would go around and around faster and faster, until they reached the speed of light in a finite time. So if a beautiful alien in a flying saucer invites you into her time machine, step with care! You might fall into one of these trapped repeating histories of only finite duration.

As I said, these results do not depend on the Einstein equations, only on the way spacetime would have to warp in order to produce closed timelike curves in a finite region. However, one can now ask: What kind of matter would an advanced civilization have to use in order to warp spacetime so as to build a finite-sized time machine? Can this matter have positive energy density everywhere, as in the cosmic string spacetime I described earlier? The cosmic string spacetime did not satisfy my requirement that

the closed timelike curves appear only in a finite region. However, one might think that this was just because the cosmic strings I used were infinitely long. One might imagine one could build a finite time machine, using finite loops of cosmic string, and have the energy density positive everywhere.

I'm sorry to disappoint people like Kip who want to return to the past, but it can't be done with positive energy density everywhere! I can prove that to build a finite time machine, you need negative energy.

The classical energy-momentum tensors of all physically reasonable fields obey the "weak energy condition," that the energy

Weak Energy Condition

The energy density is greater than or equal
to zero for all observers

density in any frame is greater than or equal to zero. Thus time machines of finite size are ruled out in the purely classical theory. However, the situation is different in the semiclassical theory, in which one considers quantum fields on a classical spacetime background. The uncertainty principle of quantum theory means that fields are always fluctuating up and down, even in apparently empty space. These quantum fluctuations would make the energy density infinite. Thus one has to subtract an infinite quantity in order to make the theory get the finite energy density that is observed. Otherwise, the energy density would curve spacetime up into a single point. This subtraction can leave the "expectation value" of the energy negative, at least locally. Even in flat space one can find quantum states in which the expectation value of the energy density is negative locally, although the integrated total energy is positive.

One might wonder whether these negative expectation values actually cause spacetime to warp in the appropriate way, the way that could lead to time machines. But it seems they must. Just before I first visited Caltech in 1974, I discovered that black holes ain't as black as they are painted!

Black Holes Ain't Black

The uncertainty principle of quantum theory allows particles and radiation to leak out of a black hole. This causes the black hole to lose mass and to evaporate slowly. For the horizon of the black hole to shrink in size, the energy density on the horizon must be negative, warping spacetime to make light rays diverge from each other. If the energy density were always positive, and warped spacetime always bent light rays toward each other, then the area of the horizon of a black hole could only increase with time. I first realized this when I was getting into bed, soon after the birth of my daughter. I won't say how long ago that was, but I now have a grandson—

The evaporation of black holes shows that the quantum energy-momentum tensor of matter can sometimes warp space-time in the direction that would be needed to build a time machine. One might imagine therefore that some very advanced civilization could arrange for the expectation value of the energy

Two-Point Function

$<\varphi(x)\varphi(y)>$ is infinite when
$x-y$ or
x and y lie on the same light ray

density to be sufficiently negative to form a time machine that could be used by macroscopic objects. However, there's an important difference between a black hole horizon and the horizon in a time machine, which contains closed light rays that keep going around and around. The energy-momentum tensor of a quantum field in a curved space background can be determined from what is called the "two-point function."

This function describes the correlations in the quantum fluctuations of the field at two points, x and y. One takes the variation of the two-point function with the positions of x and y, and then let x tend to y. The two-point function diverges as x approaches y, but one subtracts out the divergences that would occur in flat space, and those that are characterized by the local curvature at y. In curved spacetimes without closed light rays, this subtraction procedure makes the energy-momentum tensor finite, although possibly negative, as I said earlier.

However, the two-point function is also infinite if x and y can be joined by a light ray. So if there is a closed or almost-closed light ray, one has an extra infinity that is not subtracted out by the local counter terms. One would therefore expect the energy-momentum tensor to be infinite on the Cauchy horizon, the boundary of the time machine, the region in which one can travel into the past. This is borne out by explicit calculations for a few backgrounds that are simple enough for the two-point function to be known exactly. In general, the energy-momentum tensor diverges on the Cauchy horizon. In practice this would mean that a person or a space probe that tried to cross the Cauchy horizon to get into the time machine would get wiped out by a bolt of radiation!

Is this a warning from nature not to meddle with the past? In 1990, Kip and Sung-Won Kim suggested that the divergence in the energy-momentum tensor on the horizon might be cut off by

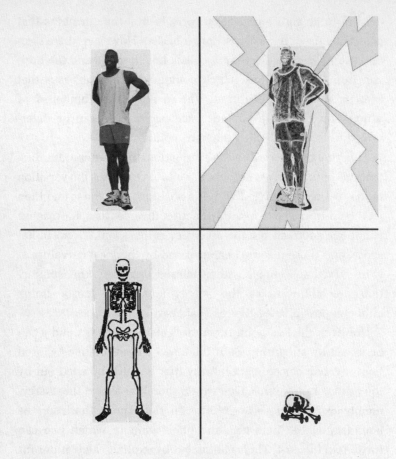

quantum gravitational effects. They argued that this could happen before the energy-momentum tensor got large enough for anyone to notice. We still don't know whether quantum gravity gives an effective cutoff, but even if it does, I think Kip would now agree that the cutoff won't come into effect in time to save any space probe from being ripped apart. So the future looks black for time travel—or should I say, blindingly white?

However, the expectation value of the energy-momentum tensor depends on the quantum state of the fields on the background.

One might speculate that there could be quantum states where the energy density was finite on the horizon, and there are examples in which this is the case. We don't know how to achieve such a quantum state, or whether it would be stable against objects crossing the horizon. But it might be within the capabilities of an advanced civilization. Whether it can be is a question that physicists should be free to discuss, without being laughed to scorn.

> Even if it turns out that time
> travel is impossible,
> it is important that we understand
> why it is impossible

To give definitive answers about quantum states at the horizon, we need to consider quantum fluctuations of the spacetime metric as well as of the matter fields. One might expect these fluctuations to cause a certain fuzziness in the light cone, and in the whole concept of time ordering. Indeed, one can regard the radiation from black holes as leaking out because quantum fluctuations of the metric mean that the horizon is not exactly defined. Because we don't yet have a complete theory of quantum gravity, it is difficult to say what the effects of metric fluctuations will be. Nevertheless, we can hope to get some pointers from the approach of another Caltech physicist, Richard Feynman.

Apart from playing the bongo drums, Feynman's great contribution to humanity was his notion that a system doesn't have just a single history, as common sense would tell us. Rather, it has every possible history, each with its own probability amplitude. There must be a history in which the Caltech football team won the Rose Bowl, though maybe the probability is low!

Richard Feynman in front of blackboard. [Courtesy California Institute of Technology, Melanie Jackson Agency.]

In the case where the system is the whole universe, each history will be a curved spacetime with matter fields in it. Since one is supposed to sum over all possible histories, not just those that satisfy some equations of motion, the sum must include spacetimes that are warped enough for travel into the past. So the question is, Why isn't time travel happening everywhere? The answer is that time travel is indeed taking place on a microscopic scale, but we don't notice it.

If one applies the Feynman sum-over-histories idea to a particle moving in a background spacetime, one has to include histories in which the particle travels faster than light, and even
backward in time. In particular, there will be histories in which
the particle goes around and around on a closed loop in time
and space. It would be like the film *Groundhog Day*, in which a
reporter has to live the same day, over and over again.

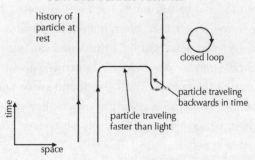

Sum over Particle Histories

One cannot observe particles with such closed-loop histories
directly with a particle detector. However, their indirect effects
have been measured in a number of experiments. One is a small

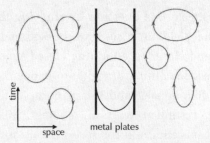

Casimir Effect

shift in the light given out by hydrogen atoms, caused by electrons moving in closed loops. Another is the Casimir effect, a

small force between parallel metal plates caused by the fact that there are slightly fewer closed-loop histories that can fit between the plates compared with the region outside.

Thus the existence of closed-loop histories is confirmed by experiment!

One might dispute whether closed-loop particle histories have anything to do with the warping of spacetime, because they occur even in fixed backgrounds such as flat space. But in recent years we have found that phenomena in physics often have dual, equally valid descriptions. I think one can equally well say that a particle moves on a closed loop in a given fixed background, or that the particle stays fixed, and space and time fluctuate around it. It is just a question of whether you do the sum over particle paths first and then the sum over curved spacetimes, or vice versa.

It seems, therefore, that quantum theory allows time travel on a microscopic scale. However, this is not of much use for science-fiction purposes, like going back and killing your grandfather. The question therefore is, Can the probability in the sum over histories have a peak near spacetimes with macroscopic closed timelike curves?

One can investigate this question by studying the sum over histories of matter fields in a series of background spacetimes that get closer and closer to admitting closed timelike curves. One would expect something dramatic to happen when closed timelike curves first appear, and this is borne out in a simple example, which I studied with my student, Mike Cassidy.[1] The background spacetimes in the series were closely related to what is called the Einstein universe. This is a static spacetime, in which time runs from the infinite past to the infinite future. The space directions, however, are finite, and they close on themselves, like the surface of Earth, but in all three dimensions. Thus the spacetime is like a cylinder, the long axis being the time direction, and the cross section being the three space

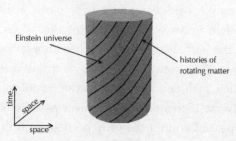

Rotating Einstein Universe
(based on work by Cassidy and Hawking)

Einstein universe

histories of
rotating matter

time

space

space

directions. Because it is not expanding, the Einstein universe does not represent the universe we live in. Nevertheless, it is a convenient background for discussing time travel, because it is simple enough that one can do the sum over histories.

Forgetting about time travel for the moment, one can consider quantum fields at finite temperature in an Einstein universe that

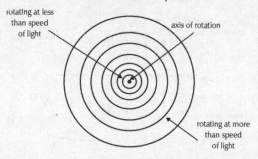

Rotation in Flat Space

rotating at less
than speed
of light

axis of rotation

rotating at more
than speed
of light

is rotating about some axis. If one were on the axis, one could remain at the same point of space. But if one were not on the axis, one would be moving through space while rotating about

the axis. If the universe were infinite in space, points sufficiently far from the axis would have to be rotating faster than
light. However, because the Einstein universe is finite in the
space directions, there is a critical rate of rotation below which
no part of the universe is rotating faster than light.

One can now consider the sum over particle histories in a
rotating Einstein universe. When the rotation is slow, there are
many paths the particle can have, each with a given amount of
particle energy. Thus many paths make contributions in the sum
over all particle histories. However, as the rate of rotation of the
Einstein universe increases, the sum over particle histories will
be strongly peaked around the only particle path that is classically allowed, namely one that is moving at the speed of light.
This means that the net sum over particle histories will be
small. Thus the probability of these backgrounds will be relatively low in the sum over all curved spacetime histories.

What do rotating Einstein universes have to do with time travel
and closed timelike curves? The answer is that they are mathematically equivalent to other backgrounds that do admit closed

Backgrounds with Closed Timelike Curves

universe expanding
in this direction

universe not
expanding in
this direction

identify with
a boost in
vertical speed

timelike curves. These other backgrounds are universes that are
expanding in two space directions but not in the third space
direction, which is periodic. That is to say, if one goes a certain

distance in this direction, one gets back to where one started. However, each time one does a circuit of the third space direction, one's speed in the first or second directions gets boosted.

If the boost is small, then there are no closed timelike curves. However, one could consider a sequence of backgrounds with increasing boosts. At a certain critical boost, closed timelike curves would appear. Not surprisingly, this critical boost corresponds to the critical rate of rotation of the equivalent Einstein universe. Since the sum-over-histories calculations in these

---

Chronology Protection Conjecture

The laws of physics conspire to prevent time travel by macroscopic objects

---

backgrounds are mathematically equivalent, one can conclude that the probability of these backgrounds goes to zero as they approach the warping needed for closed timelike curves. This supports the "chronology protection conjecture," that the laws of physics conspire to prevent time travel by macroscopic objects.

Probability Kip Could Go Back and Kill His Grandfather

Less than
1 in 10
with a
trillion trillion trillion trillion trillion
0's
after
it

Although closed timelike curves are allowed by the sum over histories, the probabilities are extremely small. In fact, based on the duality arguments I mentioned earlier, I estimate that the probability that Kip could go back and kill his grandfather is less than 1 in 10 with a trillion trillion trillion trillion trillion zeros after it. That's a pretty small probability, but if you look at Kip closely, you may see a slight fuzziness around the edges! This corresponds to the faint possibility that some bastard from the future

came back and killed his grandfather, so he's not really there!

As gambling men, Kip and I would bet on odds like that. The trouble is, we can't bet each other, because we are now both on the same side. On the other hand, I wouldn't take a bet with anyone else. They might be from the future, and know that time travel works!

Note

1. M. J. Cassidy and S. W. Hawking, "Models for Chronology Selection," *Physical Review* D57 (1998): 2372–80.

SPACETIME WARPS AND THE QUANTUM WORLD: Speculations About the Future

Kip S. Thorne

I've just been through an overwhelming birthday celebration. There are two dangers in such celebrations, my friend Jim Hartle tells me. The first is that your friends will embarrass you by exaggerating your achievements. The second is that they won't exaggerate. Fortunately, my friends exaggerated.

To the extent that there are kernels of truth in their exaggerations, many of those kernels were planted by John Wheeler. John was my mentor in writing, mentoring, and research. He began as my Ph.D. thesis advisor at Princeton University nearly forty years ago and then became a close friend, a collaborator in writing two books, and a lifelong inspiration. My sixtieth birthday celebration reminds me so much of our celebration of Johnnie's sixtieth, thirty years ago.

As I look back on my four decades of life in physics, I'm struck by the enormous changes in our understanding of the universe. What further discoveries will the next four decades bring? Today I will speculate on some of the big discoveries in those fields of physics in which I've been working. My predictions may look silly in hindsight, forty years hence. But I've never minded looking silly, and predictions can stimulate research. Imagine hordes of youths setting out to prove me wrong!

I'll begin by reminding you about the foundations for the fields in which I have been working. I work, in part, on the general theory of relativity. Relativity was the first twentieth-century revolution in our understanding of the laws that govern the universe, the laws of physics. That first revolution was brought to us by Albert Einstein in two steps: special relativity in 1905 and general relativity in 1915, with a ten-year struggle in between, much like the intellectual struggle that Alan Lightman describes in this volume.

At the end of his struggle, Einstein concluded that space and

Figure 1. Albert Einstein at age 26, when he was formulating special relativity—the first step in the first twentieth-century revolution in our understanding of the laws of nature. [Courtesy Albert Einstein Archives, the Hebrew University of Jerusalem, Israel.]

time are warped by matter and energy and that this warpage is responsible for the gravity that holds us to the surface of Earth. He gave us a set of equations from which one can deduce the warpage of time and space around the cosmic objects that inhabit our universe. In the 85 years since then, thousands of physicists have struggled with Einstein's equations, trying to extract their predictions about spacetime warpage.

In my book *Black Holes and Time Warps: Einstein's Outrageous Legacy*, I tell the story of that struggle, including the most interesting discovery it produced: the prediction of black holes. Robert Oppenheimer, shuttling back and forth between UC Berkeley and Caltech in the late 1930s, made the first, tentative form of the prediction, but it took the concerted efforts of hundreds of other physicists, in the 1950s, 1960s, and 1970s, to smoke out the full details of what a black hole is and how it should behave. My mentor, John Wheeler, was the modern pioneer of black holes, and my friend Stephen Hawking, the latter-day prophet.

A black hole is the ultimate in spacetime warpage, according to Einstein's equations: it is made wholly and solely from that warpage. Its enormous warpage is produced by an enormous amount of highly compacted energy—energy that resides not in matter but in the warpage itself. Warpage begets warpage without the aid of matter. That is the essence of a black hole.

If I had a black hole the size of the world's largest pumpkin, about 10 meters in circumference, then knowing Euclid's laws of geometry, you might expect its diameter to be 10 meters divided by $\pi = 3.14159\ldots$, or about 3 meters. But the hole's diameter is far larger than 3 meters, perhaps more like 300 meters. How can this be? Quite simply: Euclid's laws fail in the hole's highly warped space.

Consider a simple analogy. Take a rubber sheet—a child's rub-

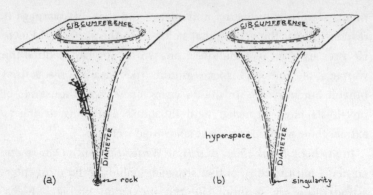

Figure 2. (a) A child's trampoline warped by a heavy rock is being explored by an intelligent ant. (b) The warped space of a black hole as seen by a hyperbeing living in hyperspace.

ber trampoline. Stretch it out between the tops of four high poles, then place a heavy rock at its center. The rock will bend the rubber downward, as shown in Figure 2a. Now suppose you are an ant living on the rubber sheet. The sheet is your whole universe. Not only are you an ant; you are a blind ant, so you can't see with your eyes the poles and rock that warp the sheet. But being a smart and inquisitive ant, you set out to explore your universe. You march around the circular edge of the sheet's indentation, pacing off its circumference: 30 meters, you conclude. Being schooled in the mathematics of Euclid, you predict a diameter of about 10 meters, but being also a skeptic of all prognostications, you set out to measure the diameter. You march inward toward the center; you march and march and march, and ultimately you come out on the other side after 300 meters of travel, not the 10 meters predicted by Euclid. "The space of my universe is warped," you conclude—highly warped.

This story is a rather accurate depiction of a black hole. We can think of the three-dimensional space inside and around a black hole as warped in a higher-dimensional flat space (often called "hyperspace"), just as the two-dimensional rubber sheet is warped as depicted in Figure 2a. If I were a higher-dimensional "hyperbeing" living in hyperspace, I would see the black hole's space to have a form very much like that of the rubber sheet (see Figure 2b).

The most intriguing thing about black holes is that, if I fall into one, there is no way for me ever to get back out nor to send signals to you as you await me on the outside. This is illustrated in Figure 3a by a two-dimensional Kip falling into a black hole

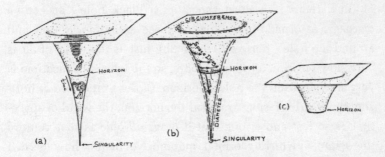

Figure 3. (a) Kip falling into a black hole and trying to transmit microwave signals to you on the outside. (b) The warpage of space and time, and the dragging of space into a tornado-like motion around a spinning black hole. (c) The warpage of space outside the horizon—the diagram serves as the template for several of the figures that follow.

as seen by a hyperbeing in hyperspace. (I have suppressed one of our universe's three dimensions to make the picture understandable.) As I fall in, I carry with myself a microwave antenna that transmits signals to you on the outside, telling what I see.

Now, not only is the space through which I move warped, so is the time, according to Einstein's equations: the flow of time slows near the hole, and at a point of no return (called the hole's "horizon," or edge), time becomes so highly warped that it starts to flow in a direction that normally would be spatial; the future flow of time is toward the hole's center. Nothing can move backward in time,[1] Einstein's equations insist; so once inside the hole, I and my microwave signals are drawn willy-nilly downward with the flow of time toward a "singularity" that lurks at the hole's core. You, waiting on the outside, can never receive my signals from beneath the horizon. They are caught by the flow of time and dragged away from you. I have paid the ultimate price in exploring the hole's interior: I can't publish my discoveries.

Besides the bending of space and the slowing and downflow of time, there is a third aspect of a black hole's spacetime warpage: a tornado-like whirl of space and time around and around the hole's horizon (Figure 3b). Just as the whirl of air is very slow far from a tornado's core, so the whirl of spacetime is very slow far from the hole's horizon. Closer to the core or horizon, the whirl is faster. And near the horizon, the whirl of spacetime is so fast and strong that it drags *all* objects that venture there into a whirling orbital motion. No matter how hard a spaceship may blast its engines, once near the horizon it cannot resist the whirl. It is dragged, by the forward flow of time, around and around inexorably—and once inside the horizon, it is also dragged downward by the forward flow of time, toward the gaping singularity at the hole's core.

The whirl of spacetime around a black hole was discovered in 1963, buried in the mathematics of Einstein's equations, by Roy Kerr, a mathematical physicist from Christchurch, New Zealand. Just as the bending of space and warping of time are produced by the hole's huge energy (the energy of the bending and warping themselves), so also the whirl of spacetime is produced by the hole's huge rotational angular momentum (an angular momentum that resides in the spacetime whirl itself). The warpage's energy and angular momentum create the warpage, according to Einstein's equations. Warpage begets warpage.

Because we can't see inside a black hole from the exterior, I will ignore its interior for awhile. I will cut off my pictures of holes at the horizon and just depict the holes' exteriors, as in Figure 3c.

Now, we relativity physicists have been terribly frustrated for the past quarter century. By 1975 we had fully smoked these black hole predictions out of Einstein's equations and were turning to astronomers for observational confirmation or refutation. But since then, despite enormous effort, astronomers have failed to produce quantitative measurements of any black hole's spacetime warpage. Their great triumphs have been a number of near-incontrovertible discoveries of black holes in the universe, but they have been unable to map, even crudely, the spacetime warpage around any of their discovered holes.

With this background in hand, I'm ready to start prognosticating. I'll begin with a prediction in which I have great confidence.

Prediction 1: In 2010 to 2015, a space-based gravitational-wave detector named LISA (Laser Interferometer Space Antenna) will reveal the warpage of spacetime around many massive black holes in the distant universe, and will map that warpage in exquisite detail—all three aspects of the warpage: the bending of space, the warping of time, and the whirl of spacetime around the horizon.

These black hole maps, each a picture of what the hole would look like as seen by a hyperbeing in hyperspace, will complete the transformation of black holes from purely theoretical entities to objects for observational exploration.

Figures 4 and 5 depict the foundation for LISA's maps. Suppose a small black hole is orbiting around a far larger black hole in the distant universe (Figure 4a). The small hole might weigh ten times as much as the Sun and have a circumference of about 180 kilometers (the size of San Francisco). The large hole

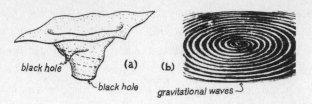

black hole   (a)   (b)

black hole   gravitational waves

Figure 4. (a) A small black hole orbiting a large black hole. (b) The gravitational waves produced by the small hole's orbiting motion.

might weigh the equivalent of a million Suns and have a circumference of about 18 million kilometers (four times larger than the Sun). The small hole would fly around the large hole at roughly half the speed of light, in an orbit only a few times bigger than the large hole's horizon.

The small hole orbiting around and around the large hole is sort of like you stirring your finger around and around in a pond of water. Just as your finger creates ripples on the water's surface that flow outward across the pond carrying information about your finger's motion, so the spacetime warpage of the small, fast-flying hole creates ripples of warpage in the fabric of spacetime around the large hole. With each complete circuit around the large hole, the small hole produces two complete oscillations of the outgoing ripples: two crests and two troughs. The ripples, called "gravitational waves" (Figure 4b), propagate out into the universe at the same speed as light. A few years ago, Fintan Ryan, a graduate student I was mentoring, showed that these waves carry, encoded in their "waveforms," a detailed map of the large hole's spacetime warpage, which is being explored by the small hole as it orbits.

These gravitational waves travel across the far reaches of intergalactic space, billions of light-years. Ultimately they reach and enter our Milky Way galaxy, and then our solar system, where they buffet LISA (Figure 5). LISA is designed to monitor

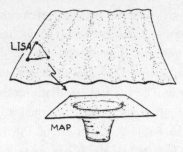

Figure 5. After traveling across the great reaches of intergalactic space, the gravitational waves buffet LISA. LISA monitors and records the waves' waveforms, and from the recorded waveforms we extract a map of the large black hole's spacetime warpage.

the waves' ripples as they go by and to record their full details. From those details we expect to decode the map that the waves carry—the map of all three aspects of the large hole's warpage.

The principle on which LISA is based is depicted in Figure 6. Two spacecraft, floating in interplanetary space, are analogous to

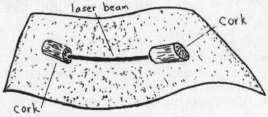

Figure 6. Just as water waves on a pond can be monitored by using a laser beam to measure the distance between two bobbing corks, so LISA will monitor gravitational waves by using a laser beam to measure the distance between spacecraft.

two corks floating on the surface of a water pond. As water waves go by, their crests and troughs stretch and squeeze the distance between the corks. The corks' relative motion can be monitored with high precision using the same technique as surveyors use: the round-trip travel of a laser beam.

Similarly, the gravitational waves stretch and squeeze space as they pass, making LISA's spacecraft move back and forth relative to each other, and that relative motion is monitored by laser beams. The greater the separation L between the spacecraft, the greater will be the tiny oscillations, ΔL, in their separation. The oscillating ratio $\Delta L/L$ is equal to the oscillating gravitational-wave field. The pattern of the oscillations as a function of time t, $\Delta L(t)/L$, is the field's gravitational waveform. This waveform is analogous to the patterns that sound waves produce when displayed on an oscilloscope, and it carries the map of the big black hole.

Figure 7 shows how one aspect of that map—the tornado-like whirl of space around the large hole—is encoded in the waveform. The space whirl drags the small hole's orbit with it, caus-

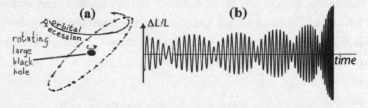

Figure 7. (a) The spin of the large black hole drags space into motion around itself, thereby causing the orbit of the small hole to precess. (b) The gravitational waves from the orbiting, precessing small hole produce tiny oscillations in the difference ΔL between the lengths of two of LISA's arms. This graph shows $\Delta L/L$ as a function of time. Each trip of the small hole around the large hole produces two oscillations in ΔL; the precession of the orbit causes a modulation of the oscillations' amplitude and phase.

ing the orbit to precess. As seen from Earth (if we could see that far with our eyes), the orbit alternates between being edge-on and approximately face-on. Correspondingly, the amplitude of the waves' oscillations (two oscillations per orbital circuit) is driven alternately smaller and then larger, so that the waves are modulated as shown in Figure 7(b). With two edge-on events in each full precession, the waveform is modulated twice as fast as space whirls.

Assume, for simplicity, that the orbit is circular and only slightly inclined to the large hole's equator, that the small hole weighs 10 Suns, and that the large hole spins very rapidly[2] and weighs a million Suns. Then one year before the small hole plunges through the large hole's horizon, its orbital circumference is just 3.4 times bigger than the horizon and there are 92,000 orbits (184,000 wave cycles) left until plunge. The waves' oscillation period is 4.8 minutes, from which we infer an orbital period (as measured by Earth-based clocks) of 2 × 4.8 minutes. And the waveform modulation period is 42 minutes *from which we infer that at 3.4 horizon circumferences, the space-whirl period is 2 × 42 = 84 minutes.*

One month before plunge, the orbital circumference is just 1.65 times larger than the horizon, the waves' oscillation period is 1.6 minutes, and there are 40,000 wave cycles left until plunge. The waveform modulation period is 8.6 minutes, *from which we infer a space-whirl period of 17.2 minutes at 1.65 horizon circumferences.*

One day before plunge, the orbital circumference is 1.028 times larger than the horizon, the wave period is 38 seconds and there are 2,000 wave cycles left. The observed modulation period is 43 seconds, *so the period of the space whirl at 1.028 horizon circumferences is 2 minutes.*

In this manner, from the waveform's changing modulation pattern, we can map the rate of space whirl as a function of location outside the horizon. With 184,000 cycles of waves to work with in the last year of the small hole's life, all coming from a region 5.8 times the size of the large hole's horizon, we expect to achieve an exquisitely accurate map.

LISA will consist of three laser-linked spacecraft residing at the corners of an equilateral triangle (Figure 8). By a variant of laser interferometry (a method explained later in the essay), the differences in the lengths of the triangle's three arms will be

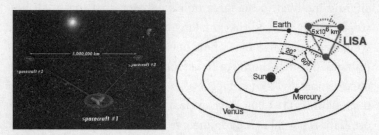

Figure 8. *Left*: LISA will consist of three laser-linked spacecraft at the corners of an equilateral triangle, 5 million kilometers on a side. *Right*: LISA's size is exaggerated here by a factor of about 10 relative to the planetary orbits.

monitored, and from the two independent arm-length differences we will deduce the waves' two independent waveforms. To extract the full map and simultaneously learn the small hole's mass and spin, the orbit's details, the large hole's orientation in space, and the distance from the holes to Earth, we must monitor both waveforms, not just one.

The distances L between LISA's three spacecraft will be 5 million kilometers (13 times larger than the Earth–Moon separation). They will travel around the Sun in the same orbit as Earth, but following Earth by about 20 degrees (50 million kilometers). After traveling across the great reaches of intergalactic space, the gravitational waves have become very weak: $\Delta L/L$ will be a little less than 10^{-21}—one part in a billion trillion. Correspondingly, the tiny oscillations ΔL in the spacecraft separations are about 10^{-10} centimeters, which is one millionth the wavelength of the laser light used to monitor the oscillations, and one hundredth the diameter of an atom. Our ability to measure such tiny motions is a tribute to modern technology!

LISA will be built and operated jointly by NASA and the European Space Agency (ESA) and is tentatively planned for launch in 2010. It was conceived (though not with this name) in the mid-1970s by several of my physicist friends: Peter Bender of the University of Colorado, Ronald Drever of Glasgow University, and Rainer Weiss of the Massachusetts Institute of Technology. Many physicists have worked hard over these past 25 years to perfect LISA's design, to figure out what kinds of wave-emitting objects it should see and what science can be extracted from their waves, and to convince NASA and ESA that LISA should be flown. At last, in the past year, LISA has won the endorsement of politically powerful committees of scientists and now seems on a fast track toward fulfilling my first prediction: exquisitely accurate maps of huge black holes in the 2010–2015 time frame.

I will turn, now, to my second prediction.

Prediction 2: Sometime between 2002 and 2008 (in other words, before LISA's 2010 launch), Earth-based gravitational-wave detectors will watch black holes collide and watch their collisions trigger wild vibrations of spacetime warpage. By comparing the observed waves with supercomputer simulations, we will discover how the warpage behaves when it interacts with itself dynamically and nonlinearly.

When water waves become so high that they interact with themselves dynamically and nonlinearly, the result can be the breaking, crashing froth that topples and engulfs surfers—or it can be an enormous tidal wave that travels across oceans at high speed, hits shores, and wreaks havoc. The analogous nonlinear, dynamical behavior of spacetime warpage is largely a mystery today. By combined gravitational-wave observations and super-computer simulations we hope to discover it.

The vehicle for our discovery is a collision between two black holes in the distant universe. The two black holes initially orbit each other and "inspiral" (gradually decrease the orbital radius) due to the loss of energy in outgoing gravitational waves. The two holes then merge in a "collision" to form a single final black hole. Lastly, the final black hole undergoes "ringdown," oscillations of decaying amplitude.

As depicted in Figure 9, each hole is like a tornado. Spacetime whirls around its horizon like air whirling around a tornado's core. As the holes orbit each other, their huge orbital angular momentum also drags spacetime into a whirling motion, so we have two tornados embedded in a third, larger tornado and crash-

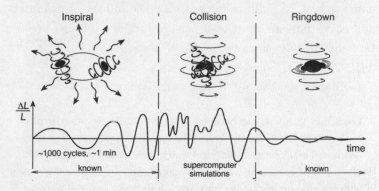

Figure 9. *Upper part:* The inspiral and collision of two black holes orbiting each other. *Lower part:* A schematic graph of the gravitational waveform emitted by the holes.

ing together. We want to know what happens when the tornados are made not from whirling air, but from a whirling spacetime warpage. To learn the answer will require a three-pronged attack: supercomputer simulations, gravitational-wave observations, and detailed comparison of the simulations and observations.

The simulations are being pursued by about fifty scientists in Europe, the United States, and Japan. These scientists are called "numerical relativists" because they are attempting to solve

Einstein's general relativity equations numerically, on comput-
ers. I have bet these numerical relativists that gravitational
waves will be detected from black hole collisions before their
computations are sophisticated enough to simulate them. I
expect to win, but hope to lose, because the simulation results
are crucial to interpreting the observed waves.

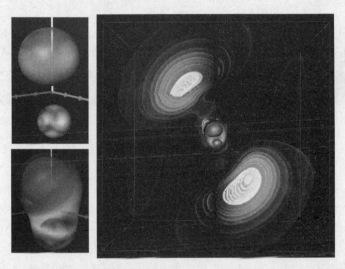

Figure 10. Simulation of the glancing, but nearly head-on collision of two
black holes with different sizes, as computed numerically on a supercomputer
by a group at the Albert Einstein Institute in Golm, Germany, led by Edward
Seidel and Bernd Brügmann. *Upper left*: Apparent horizons (close approximations to the true horizons) of the two holes shortly before the collision. *Lower
left*: Apparent horizon of the merged hole shortly after the collision, with the
individual apparent horizons inside. *Right*: Double-lobed gravitational wave
pattern produced by the collision, with the three apparent horizons at the center.
[Courtesy Albert Einstein Institute, Max Planck Society.]

Figure 10 is an example of the current state of the simulation
effort. It shows some features of a nearly head-on collision of
two nonspinning black holes with different sizes. Nothing star-
tling happens in *this* collision as a result of the warpage's

dynamical nonlinearities. By contrast, when the holes are rapidly spinning with random spin directions and collide from a shrinking, circular orbit (Figure 9), I expect complicated and wild vibrations of the warpage.

Figure 11. Aerial views of the LIGO gravitational-wave detectors at (*left*) Hanford, Washington, and (*right*) Livingston, Louisiana. [Courtesy LIGO Project, California Institute of Technology.]

Figure 11 shows three of the Earth-based gravitational-wave detectors that will discover the waves from black hole collisions sometime between 2002 and 2008, if my prediction is correct. These three detectors, two in a common facility at Hanford, Washington, and one at Livingston, Louisiana, make up LIGO, the Laser Interferometer Gravitational Wave Observatory. LIGO is part of an international network that includes a French-Italian detector called VIRGO, in Pisa, Italy; a British-German detector called GEO600, in Hanover, Germany; and a Japanese detector called TAMA in a suburb of Tokyo.

LIGO and its partners are the culmination of four decades of research by hundreds of dedicated scientists and engineers. LIGO itself began in 1983 as a dream by Rai Weiss at MIT, and by Ron Drever and me at Caltech, and has grown into a reality thanks to the leadership of MIT's Weiss, and Caltech's Robbie Vogt, Stan Whitcomb, and Barry Barish, the director of LIGO since the start of construction in 1994. Barish has built LIGO into a collaboration of about 350 scientists and engineers at about 25 institutions in the United States, Britain, Germany, Russia, Australia, India, and Japan. The enthusiasm, dedication, and effectiveness of this talented team is a marvel to behold. I'm counting on them to make my second prediction come true.

How will they do it? What kind of detectors have they built to see black holes collide? Each LIGO detector is similar to LISA. LISA's three spacecraft, which ride on the passing waves like corks on water, are replaced by four cylindrical mirrors that hang by wires from overhead supports (Figure 12), two in the

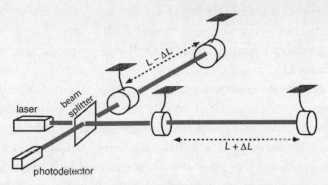

Figure 12. Schematic diagram of an Earth-based laser-interferometer gravitational-wave detector.

corner building and one each in buildings at the far ends of the **L**-shaped structures shown in Figure 11. The lengths of the **L**'s two arms are $L = 4$ kilometers. When gravitational waves fly past, oscillating much faster than the mirrors' pendular swinging frequency of 1 cycle per second, the mirrors ride the waves like a cork along horizontal directions, though the wires prevent them from riding the waves vertically. The waves' stretch and squeeze of space causes the mirrors to wiggle back and forth horizontally relative to each other, just as LISA's spacecraft wiggle back and forth. The wiggle is opposite on the detector's two arms (Figure 12), so one arm is lengthened by an amount ΔL and the other shortened by ΔL. As for LISA, the time-varying ratio $\Delta L/L$ is the gravitational waveform, and laser light is used to monitor this waveform, as follows.

Light from the laser is sent through a beam splitter (see Figure 12) so that half the light goes into each arm. The light bounces back and forth in the arms about 100 times, then emerges, and the two beams interfere with each other at the beam splitter. When one arm lengthens and the other shortens, the intensity of the light going toward the photodetector increases; when the other lengthens and the first shortens, the photodetector sees a decreased intensity. This "laser interferometry" produces a photodetector signal that is proportional to the waveform $\Delta L/L$.

LIGO's three interferometers will be fully operational by the summer of 2002, and LIGO and its international partners will then begin their first gravitational-wave search. Depending on nature's kindness, LIGO's initial sensitivity, $\Delta L/L$ about 10^{-21}, *might* or might not be good enough to observe black hole collisions. After three years of searching (and observing, we hope), LIGO's initial detectors will be replaced by "advanced detectors" with a sensitivity fifteen times better so they can look out into the universe fifteen times farther, encompassing a volume about 1,000 times greater. These advanced detectors should be able to see black hole collisions out to "cosmological distances" (a large fraction of the universe). At these distances, astrophysicists expect many collisions each year, and perhaps many each day. This estimate gives me confidence in my predictions: LIGO and its partners will begin observing black hole collisions sometime between 2002 and 2008.

I now turn from confident predictions for the present decade to an informed speculation about the decade 2020 to 2030.

Informed Speculation 3: In the decade 2020 to 2030, LIGO and its partners and a space-based successor to LISA will watch every black hole collision in the universe with hole masses below 3 million Suns, and every neutron star–black hole collision, and every neutron star–neutron star collision. They will see many collisions each day. The result, after comparing the observed waves with numerical relativity simulations, will be a huge catalog of collisions and their detailed properties, much like the catalogs of stars and galaxies produced by optical, radio, and X-ray astronomers in the twentieth century.

The "neutron stars" in this speculation are objects governed by a combination of the general relativistic laws of spacetime warpage, and the laws of quantum mechanics.

Quantum mechanics was the second great twentieth-century revolution in our understanding of physical law. Whereas the laws of spacetime warpage (the first revolution) normally act on macroscopic scales, on objects the size of a human or much larger, the laws of quantum mechanics act on microscopic scales, on objects the size of atoms or smaller. The quantum laws are as different from everyday experience as the laws of spacetime warpage, but in an even more weird way: they insist that such simple properties as the location and speed of a particle are intrinsically imprecise, and that these properties can be defined only probabilistically—a certain probability of finding the particle here, another of finding it there, and so forth. I shall discuss this weirdness shortly.

Now, quantum mechanics, among other things, governs the "nuclear force"—the force that binds neutrons and protons together in atomic nuclei. We normally probe the nuclear force in particle accelerators by slamming protons or neutrons or atomic nuclei into each other. These collision experiments have taught us many details of the nuclear force, but not all: they have taught us surprisingly little about how the nuclear force behaves when you have huge numbers of neutrons jammed together into a small volume to form bulk nuclear matter. The reason is that atomic nuclei don't get very big. They get up to a few hundred neutrons and protons in a single nucleus, but not more.

What happens when you have millions or gajillions of neutrons and protons all crammed into a tiny volume? The only place such "bulk nuclear matter" occurs in the universe today, as far as we know, is inside a neutron star, where the densities can be thirty times higher than in an atomic nucleus. So neutron stars are the key to unraveling the mysteries of bulk nuclear matter.

The quantum mechanical nuclear force determines the enormous pressure in the core of a neutron star—a pressure that tries to make the star explode. Spacetime warpage produces the enormous gravitational pull that tries to crush the neutron star, converting it into a black hole. (The enormity of the warpage is typified by the bending of space inside and around the star, as depicted in Figure 13.) Inside the star, the crushing force of grav-

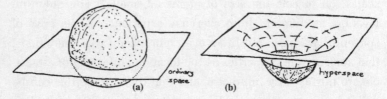

Figure 13. The warpage of space inside and around a neutron star: (a) An equatorial slice through a star, when observed from a higher-dimensional, flat hyperspace in which our universe is embedded, has the shape shown in (b). The star's circumference may be about twice its diameter rather than π times its diameter.

ity is precisely counterbalanced by the explosive force of the nuclear pressure. The star's circumference is determined by this balance: the stronger the nuclear pressure, the larger the circumference. By measuring the circumference and also the mass, we can determine the force of the star's gravity, from which we can infer the strength of the nuclear pressure—or, more precisely, we can learn about the nuclear "equation of state": the nuclear pressure as a function of density.

Although hundreds of neutron stars have been discovered with radio, optical, and X-ray telescopes, and many of their features have been probed that way, these electromagnetic observations have given us only a crude knowledge of the stars' circumferences and thence of the nuclear equation of state. The masses of about a dozen neutron stars have been measured, all coming out very near to 1.4 Suns, so they contain about 10^{57} neutrons; but their circumferences have been measured so crudely that we only know they lie somewhere between about 25 and about 50 kilometers.

This leads to my next prediction.

Prediction 4: Sometime in the years 2008 to 2010, LIGO's advanced detectors, and those of its partners, will begin probing the properties of bulk nuclear matter by monitoring the gravitational waves produced when a black hole tears a neutron star apart. The observed waves, when combined with numerical relativity simulations of the star's destruction, will tell us the star's circumference to about 10 percent accuracy. This and other features of the waves will teach us much about the nuclear equation of state.

Figure 14 depicts an example of the destruction of a neutron star by a black hole, with an accompanying emission of gravitational waves. The star and hole initially encircle each other in an orbit that gradually shrinks as it loses energy to gravitational waves. From the inspiral waves we can infer the masses and spins of the hole and star. As the star nears the hole's horizon, it encounters ever-increasing spacetime warpage, which ultimately tears the star apart. The larger the star's circumference, the more easily it gets torn, so the earlier the tearing starts. Thus

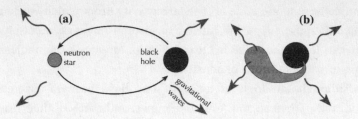

Figure 14. (a) A neutron star and a black hole that are orbiting each other
will gradually spiral inward as they lose energy to gravitational waves.
(b) As the star nears the hole, the hole's spacetime warpage may
tear the neutron star apart.

(as my graduate student Michele Vallisneri has shown), from the
onset of tearing we can infer the star's circumference and thence
some details of its equation of state; by comparing the waves
produced during the tearing with numerical-relativity simulations, we should be able to infer other equation-of-state details.

Black hole collisions and the destruction of neutron stars are
just two of many kinds of gravitational-wave sources that LISA,
LIGO, and their partners will see and will use to probe the fundamental laws of nature and their roles in the universe. But
rather than discuss others, I shall now turn to a remarkable prediction about human technology and quantum mechanics.

**Prediction 5: In LIGO in 2008, we will begin to watch 40-kilogram
sapphire cylinders behave quantum mechanically. A "quantum nondemolition technology" will be created to deal with this quantum
behavior, and already in 2008 it will be incorporated into LIGO's
advanced gravitational-wave detectors. This new technology will be
a branch of a new field of human endeavor called "quantum information science," which includes "quantum cryptography" and
"quantum computing."**

This prediction is remarkable. Textbooks say that the domain
of quantum mechanics is the microscopic world, the world of

atoms and molecules and fundamental particles. We have long known that *in principle* quantum behavior could also show up in the macroscopic world, the world of human beings, but the possibilities were so remote that we left it out of our textbooks; we hid it from our students. We must hide it no more. We must be ready, in 2008, to see the quantum mechanical "uncertainty principle" rear its head into the macro world—in LIGO's 40-kilogram mirrors—and we must learn how to evade the uncertainty principle.

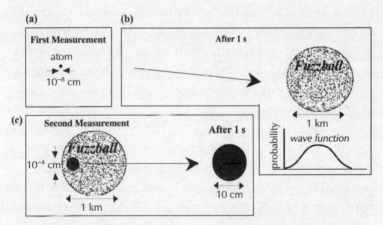

Figure 15. The uncertainty principle illustrated by successive measurements of an atom's location.

Figure 15 illustrates the uncertainty principle in the domain of atoms, where it has long held sway. Suppose we make two successive measurements of an atom's location, and in our first measurement we achieve a precision equal to the size of the atom itself, 10^{-8} centimeters (Figure 15a). The uncertainty principle says that by our very act of measuring where the atom is, we create an uncertainty in the atom's velocity. That uncertainty will cause the atom to move in an unknown and *unknowable* direction, at an unknown and *unknowable* speed. As a result, it is impossible to predict where the atom will be at the time of the second measurement. We can only say that it has a high probability of being located in some specific region, sometimes called the atom's "quantum fuzzball" (Figure 15b). The longer we wait between measurements, the larger will be the fuzzball. If we wait just 1 second, the uncertainty principle predicts a fuzzball size of *1 kilometer!* The probability for finding the atom at various points inside this 1-kilometer fuzzball is described by the atom's "wave function" (Figure 15b). The laws of quantum mechanics give us precise ways to predict this wave function—that is, the probability of where the atom is—but the exact location is unpredictable.

Suppose that, when the fuzzball has expanded to 1 kilometer, we make a second measurement of the atom's location, this time with an accuracy 10,000 times worse than the first—an accuracy of 10^{-4} centimeter. This second act of measurement suddenly shrinks the 1-kilometer fuzzball to 10^{-4} centimeter (Figure 15c) and also produces a new velocity uncertainty. According to the uncertainty principle, the velocity uncertainty is inversely proportional to the accuracy of the location measurement, so during the 1 second of time after our second measurement, the fuzzball grows to a size of 1 kilometer divided by 10,000, or just 10 centimeters (Figure 15c).

As weird as the uncertainty principle may seem, it is reality. It has been verified in many laboratory experiments. A key feature of the uncertainty principle is that the velocity uncertainty created by a location measurement is not only inversely proportional to the location accuracy; it is also inversely proportional to the mass of the measured object. That is why we have never yet seen a human-sized object behave quantum mechanically: Our huge human masses—10^{28} times larger than the mass of an atom—cause our velocity uncertainties and quantum fuzzballs to be exquisitely small.

It is a remarkable tribute to the LIGO scientists that their technology will reveal the tiny fuzzball behavior of 40-kilogram mirrors in 2008 (if my prediction is correct). Figure 16 shows the kind of mirrors I am talking about. The mirrors in the pictures are the ones for LIGO's first detectors, the detectors that will begin their gravitational-wave search in 2002. These initial mir-

Figure 16. *Left*: A mirror for the initial LIGO interferometers, resting on a velvet cushion. *Right*: The mirror being hung by wires in its cradle in LIGO. [Courtesy LIGO Project, California Institute of Technology.]

rors weigh 11 kilograms, not 40, and are made of quartz, not sapphire, but the advanced 40-kilogram sapphire mirrors in 2008 will look very much like these.

The influence of the uncertainty principle on one of LIGO's advanced 40-kilogram sapphire mirrors is depicted in Figure 17.

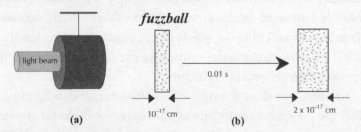

Figure 17. The consequences of the uncertainty principle for successive measurements of the center of mass of an advanced LIGO mirror.

The light beam that measures the mirror's location does so by averaging over a 10-centimeter-diameter light spot on the mirror's face, and averaging over about 1 millisecond of time—far longer than the periods of thermal vibration of the mirror's individual atoms. This averaging guarantees that the beam measures the average location of all the atoms—that is, it measures the location of the mirror's "center of mass." In effect, the mirror in this measurement behaves like a single particle weighing 40 kilograms, instead of like a conglomerate of 10^{28} atoms banging around against each other.

The light beam does not measure the center-of-mass location in all three dimensions, but in only one: along the beam's direction. In 2008 to 2010 it will measure that location with exquisite accuracy: about 10^{-17} centimeters—1/10,000 the diameter of an atomic nucleus, 1 billionth the diameter of an atom, 10^{-13} (one ten-trillionth) of the wavelength of light. This fantastic precision will localize the mirror's center of mass into the 10^{-17}-centimeter-thick fuzzball shown in Figure 17b. If that fuzzball did not grow between measurements, then by successive 10^{-13}-centimeter measurements, we could detect gravitational waves that move LIGO's mirrors by distances as small as $\Delta L = 2 \times 10^{-17}$ centimeters. However, the uncertainty principle forces the fuzzball to grow: The first measurement, with its extreme accuracy, induces a velocity uncertainty large enough to double the fuzzball thickness in half a gravitational wave period (about 1/100 second). This growth will hide the effects of any gravity wave as small as $\Delta L = 2 \times 10^{-17}$ centimeters—unless we can find some way to circumvent the uncertainty principle.

In 1968, my close Russian friend Vladimir Braginsky identified the uncertainty principle as a potential obstacle for gravitational-wave detectors and other distant-future, high-precision measuring devices, and in the 1970s Braginsky had the foresight to begin inventing ways to circumvent it—ways to which he gave the name "quantum nondemolition," meaning "don't let the uncertainty principle demolish the information you are trying to extract from your measuring apparatus." I and my students joined Braginsky in this quest for a few years in the late 1970s, and we recently renewed our collaboration with vigor when we realized that LIGO must confront the uncertainty principle in 2008. Thanks to ideas from Braginsky and his Russian colleagues, and to recent work by Alessandra Buonanno and Yanbei Chen in my own group, we will be ready in 2008: we now know viable ways to protect the gravitational waves' information from the uncertainty principle as it passes through LIGO's 40-kilogram, quantum mechanical mirrors.

The keys to this quantum nondemolition are several, and are mostly too complex to discuss here today, but one key idea can be stated quite simply: in the advanced detectors we must *never* measure the locations of the mirrors, nor the separations between them (which would contain location information). Instead, we must measure only *changes* in the separations, without ever measuring the separations themselves. In this way, we can evade the clutches of the uncertainty principle.

Early in this essay, I depicted myself futilely trying to send signals out of a black hole as I was pulled into the singularity at its core (Figure 3a). The nature of that singularity is a great mystery, but the spacetime warpage near it is not. In the early 1970s, three Russian friends of mine, Vladimir Belinsky, Isaac Khalatnikov, and Yevgeny Lifshitz, probed the singularity's warpage by solving Einstein's equations, and discovered the violent and chaotic behavior shown in Figure 18.[3] When I near the singularity, the warpage stretches me from head to foot while squeezing my sides, then stretches my sides while squeezing me

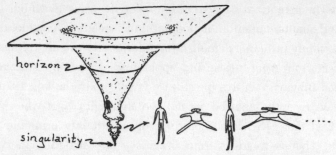

Figure 18. Kip falls into a black hole. As he approaches the singularity at its core, he is chaotically stretched and squeezed by the extreme spacetime warpage there.

from head to foot, and then repeats, faster and faster, in an ever-changing, chaotic pattern. Soon my body gives way, and I become "spaghettified" (to use a technical term coined by John Wheeler). Then my body's individual atoms get spaghettified beyond all recognition, and then space itself gets spaghettified.

I am convinced, by arguments given by Wheeler in 1957, that the end point of spaghettification—the singularity itself—is governed by a union, or marriage, of the laws of quantum mechanics and those of spacetime warpage. This must be so, since the warpage spaghettifies space on scales so extremely microscopic that they are profoundly influenced by the uncertainty principle.

The unified laws of spacetime warpage and quantum mechanics are called the "laws of quantum gravity," and they have been a holy grail for physicists ever since the 1950s. In the early sixties, when I was a student of Wheeler's, I thought the laws of quantum gravity so difficult to comprehend that we would never discover them in my lifetime, but I now am convinced otherwise. An approach to understanding them, called "string theory," looks tremendously promising.

String theory has a bad reputation in some circles because it has not yet made predictions that are testable in the laboratory or by astronomical or cosmological observations. Singularities, being quantum gravitational objects, may rectify this, if we can observe them.

Now, the singularities inside black holes are not of much use, since they can't be seen from Earth. If you were to see them, you would die without publishing. Are there other singularities that we *can* observe without dying? Yes, there is at least one: The big bang singularity that gave birth to our universe, and gravitational waves are the ideal tool for probing it.

The big bang produced three kinds of radiation: electromagnetic radiation (photons), neutrino radiation (neutrinos), and gravitational waves (Figure 19). During the first 100,000 years of its life, the universe was so hot and dense that photons could not propagate; they were created, scattered, and absorbed before

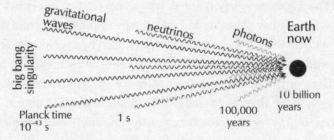

Figure 19. Photons, neutrinos, and gravitational waves from the big bang creation of the universe.

traveling hardly any distance at all. Finally, at age 100,000 years, the universe had expanded and cooled enough for photons to survive, and they began their journey to Earth. We see them today as a "cosmic microwave background" (CMB), arriving from all directions and carrying a picture of the universe at age 100,000 years.

Neutrinos are much more penetrating than photons. Someday the technology of neutrino detectors will become good enough to detect and measure the neutrinos from the big bang. When that happens, they will bring us a picture of the universe at age 1 second; before then, the universe was too hot and dense for neutrinos to survive.

Gravitational waves are far more penetrating than neutrinos—so penetrating, according to calculations by my Russian friends Yakov Borisovich Zel'dovich and Igor Novikov, that they should never have been absorbed or scattered by the universe's matter. They should have traveled unscathed by matter, from the universe's earliest moments—from the big bang singularity itself. They therefore might bring us a picture of the universe's birth throes—birth throes which, over a time of about 10^{-23} second (known as the "Planck time"), destroyed the singularity and created space, time, matter, and radiation.

The big bang's gravitational waves, no matter how weak and few, should have been amplified strongly in the first second of the universe's life. This amplification (predicted by my Russian friend Leonid Grishchuk in the mid-1970s) is caused by nonlinear interactions of the waves with the universe's spacetime warpage, and it gives hope that the waves will be strong enough to detect. This leads to my next prediction—really an informed speculation, since I'm less confident of it than of the things I call predictions.

Informed Speculation 6:[4] Sometime between 2008 and 2030, gravitational waves from the big bang singularity will be discovered. There will ensue an era, lasting at least until 2050, in which great efforts will be made to measure the spectrum of the primordial gravitational waves (their intensity as a function of wavelength) from wavelengths of 10 billion light-years down to 100 meters, and to map out the waves' intensity pattern on the sky. These efforts will reveal intimate details of the big bang singularity, and will thereby verify that some version of string theory is the correct quantum theory of gravity. They will also reveal a great richness of phenomena in the first second of the universe's life.

Why am I so uncertain about the date to discover waves from the big bang singularity (2008 to 2030)? Because we are extremely ignorant of the singularity's properties and the universe's first second of life. The physics establishment likes a model for the first second called "inflation," which predicts that big bang gravitational waves are so weak that to detect them may require the technology of 2030. However, I'm skeptical of this establishment prediction because its inflationary model does not take detailed account of the (as yet unknown) laws of quantum gravity. An initial attempt to incorporate string theory (our best guess at quantum gravity) into the big bang physics has been made by Gabriel Veneziano in Switzerland and others. Their stringy big bang model predicts waves that might be strong enough for detection by LIGO in 2008 or LISA in 2010. But string theory is still in its infancy, and the stringy model is necessarily crude and tentative, so I have little confidence in its predictions. Nevertheless, these predictions are a warning that the big bang and its gravitational waves may be quite different from the establishment's pessimistic inflationary views; the big bang waves might well be detected before 2030.

The establishment also tells us with much confidence that in the universe's first second of life, there must have been a rich variety of activity. For example, as the universe expanded, it cooled down from an initial, unbelievably hot temperature. Initially, all the fundamental forces—the gravitational force, electromagnetic force, weak nuclear force, and strong nuclear force—were unified into a single force. Thereafter, at discrete moments in the expansion and cooling, each force suddenly and violently acquired its own identity, perhaps producing strong gravitational waves in the violence. For example, the electromagnetic force is predicted to have acquired its own identity last, by splitting off from the weak nuclear force when the universe's temperature was about 10^{16} degrees and its age was about 10^{-15} second (a thousandth of a trillionth of a second, also called a femtosecond). The gravitational waves produced in this "birth of the electromagnetic force" should lie within LISA's wavelength band today, and might be strong enough for LISA to detect and use to watch the birth of electromagnetism.

Though gravitational waves from the big bang singularity may be a promising way to probe the laws of quantum gravity, they are far from a sure bet. It would be much nicer if we had other singularities to probe.

Is there any hope ever to find and study singularities in the present-day universe? The establishment's answer of "probably not" is embodied in Roger Penrose's "cosmic censorship conjecture," which says that all singularities except the big bang are hidden inside black holes; that is, they are clothed by horizons. *There are no naked singularities.*

In 1991 Stephen Hawking, and John Preskill and I made a bet about cosmic censorship, displayed in Figure 20. Hawking as defender of the establishment (he has even been anointed "companion of honor to her majesty the Queen of England"!) insists that "naked singularities are . . . prohibited by the laws of physics," while Preskill and I, tweaking the establishment, maintain that naked singularities are "quantum objects that may exist unclothed by horizons, for all the Universe to see."

Now, Preskill and I were far from confident we would win, but Hawking conceded in 1997 (Figure 21 left), though not gracefully. Our bet specified that "the loser will reward the winner with clothing to cover the winner's nakedness. The clothing is to be embroidered with a suitable concessionary message." The clothing Hawking gave us was a politically incorrect T-shirt that Preskill's wife and my wife forbade us to wear in public, but that I show here (Figure 21 right) for all the world to see. Although

Whereas Stephen W. Hawking firmly believes that naked singularities are an anathema and should be prohibited by the laws of classical physics,

And whereas John Preskill and Kip Thorne regard naked singularities as quantum gravitational objects that might exist unclothed by horizons, for all the Universe to see,

Therefore Hawking offers, and Preskill/Thorne accept, a wager with odds of 100 pounds stirling to 50 pounds stirling, that when any form of classical matter or field that is incapable of becoming singular in flat spacetime is coupled to general relativity via the classical Einstein equations, the result can never be a naked singularity.

The loser will reward the winner with clothing to cover the winner's nakedness. The clothing is to be embroidered with a suitable concessionary message.

Stephen W. Hawking   John P. Preskill & Kip S. Thorne
Pasadena, California, 24 September 1991

Figure 20. The 1991 bet in which Hawking upholds the cosmic censorship conjecture, and Preskill and Thorne oppose it. [From *Black Holes and Time Warps: Einstein's Outrageous Legacy* by Kip S. Thorne. Copyright © 1994 by Kip S. Thorne. Used by permission of W. W. Norton & Company, Inc.]

Figure 21. *Left*: Hawking concedes he has lost our cosmic censorship bet, as Thorne bows with pleasure and Preskill looks on with glee. *Right*: The politically incorrect T-shirt that Hawking gave us. [Photo at left taken at Caltech, courtesy Irene Fertik, University of Southern California.]

Hawking was conceding that the laws of physics permit naked singularities, the message on the T-shirt insists—as Hawking still does—that "Nature abhors a Naked Singularity." This was hardly a "suitable concessionary message."

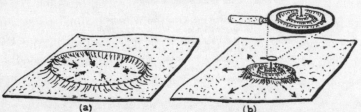

(a)   **(b)**

Figure 22. A sketch of the supercomputer simulations of imploding waves that triggered Hawking to concede that the laws of physics permit naked singularities, at least in principle.

To explain Hawking's insistence, I sketch in Figure 22 the evidence that triggered his concession. That evidence came from supercomputer simulations of an imploding, spherical pulse of waves (Figure 22a). The original simulations, by Matthew Choptuik at the University of Texas, were a tour de force in numerical relativity—far more accurate than any previous numerical relativity computations—but they involved a simple type of wave that might not exist in the universe: a "classical scalar wave." Later simulations, by Andrew Abrahams and Chuck Evans at the University of North Carolina, involved imploding gravitational waves and gave the same results.

When the imploding waves were given a big amplitude, so they contained lots of energy, the implosion's dynamical nonlinearities produced a singularity clothed by a black hole horizon, as all gravitational physicists had expected. When the waves were given a small amplitude, so that they contained only a little energy, the waves went inward, passed through each other unscathed by any nonlinearities, and reemerged as outgoing waves. This was also expected.

The big surprise came when the waves' amplitude was care-fully tuned to infinitesimally less than enough to make a black hole. Then, as depicted in Figure 22b, the imploding waves inter-acted with each other dynamically and nonlinearly to produce a boiling froth of spacetime warpage from which outgoing waves continually leaked. Close examination of the boiling at the core showed the waves shrinking in wavelength—continually, quickly, and with a surprisingly regular pattern—until they created an infinitesimally small, naked singularity, which (we suspect) lives for an infinitesimally short time before destroying itself.

With these simulations as a guide, we were able to look back at previous pencil-and-paper studies of Einstein's equations by Demetrios Christodolou (a former postdoc of mine, now a pro-fessor of mathematics at Princeton University) and see that they confirmed that the implosion could produce a naked singularity. It is a tribute to numerical relativity that only after numerical simulations revealed the full details of the imploding waves' froth were we able to understand clearly what Christodolou's mathematics was trying to say. What a wonderful tool comput-ers have become, thanks to people like Choptuik, Abrahams, and Evans!

So why does Hawking insist that nature abhors naked singu-larities? Because, to make their singularities naked, Choptuik, Abrahams, Evans, and Christodolou had to adjust very precisely the amplitude of the imploding waves (Figure 22b). With a slightly higher amplitude, a singularity would form but would be hidden by a black hole horizon; with a slightly lower ampli-tude, the waves would interact and boil, then reexplode without making any singularity at all. Only one, very delicately chosen amplitude would produce a naked singularity, and that singular-ity would be infinitesimal in size, energy, and (presumably) life-time. Such fine-tuning of amplitudes is extremely unlikely ever to occur in nature—though highly advanced civilizations might achieve it in their laboratories. Regrettably, human civilization is utterly incapable of producing and fine-tuning the required waves—today, next year, the next century, and probably the next millennium.

Hawking, Preskill, and I are persistent in our pursuit of truth and fun, so we have renewed our bet (Figure 23). Hawking now

> **Whereas Stephen W. Hawking (having lost a previous bet on this subject by not demanding genericity) still firmly believes that naked singularities are an anathema and should be prohibited by the laws of classical physics,**
>
> **And whereas John Preskill and Kip Thorne (having won the previous bet) still regard naked singularities as quantum gravitational objects that might exist, unclothed by horizons, for all the Universe to see,**
>
> **Therefore Hawking offers, and Preskill/Thorne accept, a wager that**
> > *When any form of classical matter or field that is incapable of becoming singular in flat spacetime is coupled to general relativity via the classical Einstein equations, then*
> **A dynamical evolution from generic initial conditions** (*i.e., from an open set of initial data*) **can never produce a naked singularity** (*a past-incomplete null geodesic from* $\mathcal{I}_+$).
>
> **The loser will reward the winner with clothing to cover the winner's nakedness. The clothing is to be embroidered with a suitable, truly concessionary message.**
>
> Stephen W. Hawking   John P. Preskill & Kip S. Thorne
>
> Pasadena, California, 5 February 1997

Figure 23. The new, 1997 version of the bet, in which Hawking upholds the cosmic censorship conjecture, and Preskill and Thorne opposite it. The type in italics makes the bet more precise by using the technical jargon of theoretical physics.

insists that, if we rule out fine-tuning (in the words of the bet, for "generic initial conditions"), naked singularities cannot be made, which means they cannot arise naturally. Preskill and I again disagree, and require that next time around, the clothing be embroidered with a *truly* concessionary message.

I will venture a prediction about the outcome of our bet.

Prediction 7: Before Hawking, Preskill, and I die, our new cosmic censorship bet will get resolved. Who will win? Hawking, I fear, but it is not obvious and I won't undermine our bet by predicting. However, I do predict that the effort to resolve our bet—to find out whether naked singularities can form without fine-tuning—will involve three prongs: pencil-and-paper calculations, numerical relativity calculations, and gravitational-wave searches.

The gravitational-wave searches will be part of LISA's project to make detailed maps of the spacetime warpage around massive black holes (see Figures 5 and 7). If one or more of the maps differs from general relativity's black hole predictions, it may be that the central object, into which the small-mass hole is spiraling, is a naked singularity rather than a massive black hole. The odds of this are small, but we will have the tools to search, so search we will.

I turn, now, to my final set of predictions, all focusing on the laws of quantum gravity and what they will teach us.

Prediction 8: By 2020, physicists will understand the laws of quantum gravity, which will be found to be a variant of string theory. By 2040 we will have used those laws to produce high-confidence answers to many deep and puzzling questions, including:

- **What is the full nature of the big bang singularity, in which space, time, and the universe were born?**
- **What came before the big bang singularity, or was there even such a thing as a "before"?**
- **Are there other universes? And if so, how are they related to or connected to our own universe?**
- **What is the full nature of the singularities inside black holes?**
- **Can other universes be created in black hole singularities?**
- **Do the laws of physics permit highly advanced civilizations to create and maintain wormholes for interstellar travel, and to create time machines for backward time travel?**

Wormholes and time machines are discussed at length in the essays by Novikov and Hawking, earlier in this volume, and in the last chapter of my book *Black Holes and Time Warps*—and they are also well known to any consumer of Hollywood films and television. Figure 24 shows a lighthearted example of a wormhole, using drawings adapted from my book.

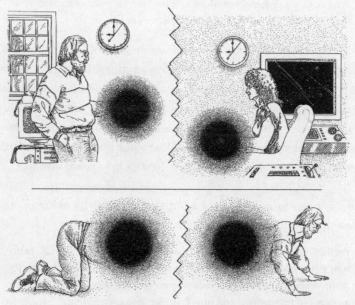

Figure 24. *Top*: Carolee travels out through the universe in a spaceship, I remain at home on Earth, and we hold hands through a wormhole. *Bottom*: I crawl through the wormhole from Earth to the spaceship. [Adapted from drawings by Matthew Zimet from *Black Holes and Time Warps: Einstein's Outrageous Legacy* by Kip S. Thorne. Copyright © 1994 by Kip S. Thorne. Used by permission of W. W. Norton & Company, Inc.]

My wife, Carolee Winstein, and one mouth of a wormhole are in a spaceship far from Earth, and I am in our home in Pasadena near the other mouth. The distance through the wormhole is very short, so Carolee and I can hold hands through it, romantically, as she sails around in interstellar space (Figure 24 top). If we want to do more than hold hands, then I can crawl through the wormhole (Figure 24 bottom) into her spaceship.

In my book I explain a crucial consequence of Einstein's laws of spacetime warpage: to hold a wormhole open so that I or anything else can travel through it, the wormhole must threaded with "exotic material"—material that, as seen by someone at rest inside the wormhole, has an enormous rubber-band-like tension; a tension larger than its enormous energy density. (I have never explored whether I could crawl through the exotic material with impunity, since we know so little about exotic material.)

We do know that such exotic material, as Hawking describes in his essay and as I discuss in my book, can actually exist in tiny quantities and under very special circumstances. However, establishment physicists strongly suspect that physical law will forbid anyone from ever concentrating enough of it together for a long enough time to hold a human-sized wormhole open. One reason for this prejudice is that someone moving through exotic material at high speed, instead of being at rest in it, will see a negative energy density. This means it violates the "weak energy condition" discussed in Hawking's essay, and the physics establishment has an amorous relationship with the weak energy condition.

In the years since I wrote my book, several of my physicist friends have worked hard trying to figure out whether the laws of physics would permit a very advanced civilization to place enough exotic material in a human-sized wormhole to hold it open. The final answer is not in, and might not be fully in until the laws of quantum gravity are fully understood. However, the tentative results, mostly from my former student Eanna Flanagan and my friends Bob Wald, Larry Ford, and Thomas Roman, look bad for wormholes.

Despite this, I remain an optimist. If pressed to speculate (as I am pressing myself today), I offer the following.

Not-So-Well-Informed Speculation 9: It will turn out that the laws of physics do allow sufficient exotic matter in wormholes of human size to hold the wormholes open. But it will also turn out that the technology for making wormholes and holding them open is unimaginably far beyond the capabilities of our human civilization.

Why am I optimistic about large amounts of exotic matter? Perhaps mostly because of my skepticism about our present understanding of what kinds of matter can exist in the universe. This skepticism is triggered by a recent cosmological discovery.

Only about 5 percent of the mass of the universe is in the kind of material from which humans are made—"baryonic matter" (molecules, atoms, protons, neutrons, electrons, and so forth). Roughly 35 percent is in some unknown form of "cold dark matter," which (like baryonic matter) can get pulled inward by gravity to form halos around galaxies, and might also form dark matter "galaxies," "stars," and "planets" that emit no light. As for the remaining 60 percent of the universe's mass: it is in some equally unknown form of "dark energy" (as cosmologists call it) that pervades the whole universe and possesses an enormous tension. [5] Is its tension larger than its energy density? Might it thereby be the kind of exotic material that is needed to hold wormholes open? We don't know for sure, but establishment physicists have a very strong prejudice that its tension is equal to its energy density, or perhaps a bit smaller, but not larger. I tend to agree: we should be so lucky as to have nature provide us with exotic material in profusion, everywhere in the universe!

Nevertheless, the dark energy gives me hope that exotic material can, in fact, exist in large amounts. Why? For the simple reason that the dark energy warns us of how very ignorant we are.

And what of time machines? In *Black Holes and Time Warps* I described a universal mechanism, identified by my postdoc Sung-Won Kim and me in 1990, that might always make a time machine self-destruct at the moment one tries to activate it. Hawking discusses this mechanism in his essay, using the fancy words: "In general, the energy-momentum tensor diverges on the Cauchy horizon." A more visual, make-believe description is shown in Figure 25.

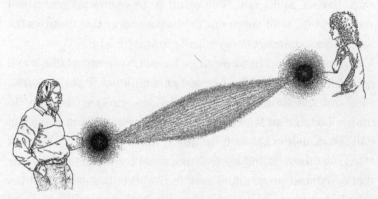

Figure 25. The self-destruction of a time machine at the moment it is first activated. [Drawing by Matthew Zimet from *Black Holes and Time Warps: Einstein's Outrageous Legacy* by Kip S. Thorne. Copyright © 1994 by Kip S. Thorne. Used by permission of W. W. Norton & Company, Inc.]

In my home in Pasadena, the flow of time is slowed a bit by Earth's mass, while in Carolee's spaceship in interstellar space, with no massive bodies nearby, time flows at its normal, faster pace. As Novikov describes in his essay, after a while this difference of flow rate transforms the wormhole into a time machine: Carolee can travel backward in time by crawling through the wormhole, and can then climb into another spaceship and fly out into interstellar space and meet her younger self.

There is a first moment, in Carolee's spaceship, when time travel becomes possible—the moment of "time machine activation." This is the moment when the fastest traveling entity of all, a bit of radiation moving at light speed, can pass through the wormhole from spaceship to Earth, then travel at light speed back to the ship through interstellar space, arriving at the same moment as it started out.[6] The result is two copies of each bit of radiation, the younger copy and the older copy, inhabiting the same spot in space and time. These two copies then travel through the wormhole and back making four copies, then eight, then sixteen, and so on. The result is an enormous number of radiation bits, with enormous explosive energy that destroys the wormhole, according to my calculations with Kim.

Our calculations, however, were based on general relativity and quantum theory in their unmarried, ununified forms. In 1990, Kim and I, scrutinizing our calculations, guessed that the ill-understood unified laws of quantum gravity would intervene and halt the explosion before it destroyed the time machine. Stephen Hawking disagreed and showed us a more cogent viewpoint—one that convinced us quantum gravity would only intervene at the very last moment, just as the time machine was on the verge of destruction. Quantum gravity, it seemed, would hold the answer tightly in its own grip. We could not know the fate of time machines until we fully understand the laws of quantum gravity.

That was the state of things in 1994, when my book was published. In the past six years, new calculations have given conflicting hints: On one hand, as Stephen says in his essay, "One might speculate that there could be quantum states [situations] where the energy density was finite on the [Cauchy] horizon [that is, where the time machine does not come close to self-destruction], and there are examples [of calculations] in which this is the case." By creating such quantum states (situations), an advanced civilization might successfully make and activate a time machine. However, these quantum states look unrealistic; I doubt they can be made in the real universe.

On the other hand, Stephen and his student Mike Cassidy have used a tentative, queasily infirm version of the laws of quantum gravity to estimate what they say about self-destruction. This version of quantum gravity predicts an exquisitely tiny probability for a time machine to escape self-destruction: 1 part in 10^{60}—one-trillionth of a trillionth of a trillionth of a trillionth of a trillionth. Can we believe this calculation? I don't know, but it is probably our best guide today to the fate of time machines.

All versions of the quantum gravity laws are queasily infirm today. But they are becoming much more firm with the march of time, and by 2020 (if my prediction 8 is right), they will be fully firm. What will they then tell us about time machines? I offer the following.

Speculation 10: It will turn out that the laws of physics forbid backward time travel, at least in the macroscopic world of human beings. No matter how hard a highly advanced civilization may try, it cannot prevent any time machine from self-destruction at the moment of activation.

Sadly, Stephen won't bet me on this. We find ourselves on the same side. He has convinced me, but only at the level of informed speculation.

So there you have them: ten speculations and predictions about the future. All ten will be proved or disproved long before my next big birthday bash, sixty years hence. The research that probes them will radically change our viewpoint on spacetime warps and the quantum world.

Notes

1. Or, more precisely, nothing can move backward through the *local* flow of time. If backward time travel is possible, then (as Novikov explains earlier in this volume) it can be achieved only by going on a round-trip (such as through a wormhole), on which you are always

moving forward with the local flow of the "river of time" but you return to where you started earlier than you set out. I shall prognosticate about this at the end of my essay.

2. For experts in the mathematics of black holes: I am assuming that a/M = 0.999.

3. More recent research, initiated by my Canadian–South African friend Werner Israel, has shown that, as the black hole ages, the warpage around its singularity becomes more tame, and possibly even less lethal. I'm skeptical that it grows less lethal, I but must admit my skepticism is not firmly founded. Only the laws of quantum gravity (discussed elsewhere in my essay) know for sure.

4. I have modified this speculation from the version in my talk at Kipfest, due to new insights in the months since then and before this book went to press.

5. In the original, oral version of this essay, I offered a prediction that by 2002 it would be completely clear that dark energy exists. I have deleted that prediction from this written version of my talk because, as this book goes to press in January 2002, new cosmological observations have confirmed its existence with high confidence. It seems much less daring now to predict that dark energy is real than it did in June 2000!

6. Actually, as is described in my book, it is "vacuum fluctuations" of the radiation field that make this first trip and pile up on themselves.

ON THE POPULARIZATION OF SCIENCE

Timothy Ferris

Science is young. The scientific enterprise has been a going concern for less than half of the 1,000 years that Alfred North Whitehead estimated are required for a new mode of thought to penetrate to the core of a culture. Yet science has already transformed much of the world, in at least three ways: technologically, intellectually, and politically.

The technological achievements of science have made the developed world wealthy and healthy, if not always wise, but have also raised its anxiety level. Part of this anxiety arises from the reasonable apprehension that technological power, like all power, has its dangers. But some of it has to do with the fact that so many people find themselves surrounded—and, sometimes, threatened—by machines whose functioning they don't understand, and behind which stands a scientific enterprise that they also don't understand.

Intellectually, science has created a new way of thinking, in which fear, superstition, and blind obedience to authority are replaced by a form of reasoned, open-minded inquiry rooted in

observation and experiment. As a result, the scientifically edu-
cated now see themselves as enmeshed in a web of life from
which they have sprung, aboard one among billions of planets
and in an expanding universe of unknown and perhaps infinite
extent. For some, this new view is exciting and stimulating, but
for others it seems vaguely threatening. They turn from the telescope to ask, "Doesn't this all make you feel insignificant?"
Perhaps the more precise word would be "insecure." Science
threatens to undermine not only old conceptions of ourselves
(such as the notion that we occupy the center of the universe)
but also old ways of thinking (for example, that our feeling
deeply that something surely must be true has any relation to
the question of whether it actually can be demonstrated to be
true). This threat is real on both counts, and should be acknowl-
edged as such by those of us who popularize science—although
we are also free, if we feel comfortable living with such dangers,
to explain why.

Third, and less widely discussed to date, is what might be
called the political contribution of science. It is no coincidence
that the publication of Isaac Newton's *Principia*, in 1687, stands
at the headwaters of the Enlightenment. Or that among the
founders of democratic movements in the American colonies
and elsewhere were to be found a disproportionately large number of scientific thinkers. Or that scientists figure prominently
among the dissident movements in totalitarian nations today.
Science is inherently antiauthoritarian: It replaces the top-down
systems of political thought that Thomas Paine lumped together
under the term "despotism" with a bottom-up approach, in
which anyone capable of making competent observations and
conducting controlled experiments may properly be regarded as
a potential source of authority—an authority that resides not in
the individual but in his or her findings.

Science encourages—indeed, requires—us to live with doubt and ambiguity, and to appreciate the enormity of our own ignorance. These habits of mind have, to some degree, percolated into the domain of political as well as scientific affairs. As Richard Feynman put it, "The government of the United States was developed under the idea that nobody knew how to make a government, or how to govern. The result is to invent a system to govern when you don't know how. And the way to arrange it is to permit a system, like we have, wherein new ideas can be developed and tried out and thrown away."[1]

In addition, the conduct of scientific research demands freedom of expression and association. It's hard enough to do physics without also being told that you cannot go to half the relevant conferences, and that your scientific ideas must comport with the enshrined philosophy of the government. This demand for freedom makes allies of scientists, writers, and artists. It also imposes a considerable handicap on nations that seek to compete in an increasingly scientific and technological world while denying freedom to their citizens. And so, in my view, while science is responsible for having invented terrifying weapons of war, it is also at least partly responsible for the fact that nearly half of all humanity now lives in democratic societies, loosely defined, and that the year 2000 dawned on a world in which no two nations were at war. Blamed for building bombs, science also fights for freedom.

To sum up: Technologically, intellectually, and even politically, science resides somewhere near the center of our culture, by which I mean the society of all those persons who value their freedom, honor their responsibilities, appreciate their ignorance, and are willing to keep learning. And yet at the same time, most of the citizens of these same societies remain estranged from science.

Each year we read stories in the newspapers about what's being called "scientific illiteracy." They tell us, for instance, that almost half of all Americans deny that human beings evolved from earlier species of animals, that a majority don't know that the solar system is located in the Milky Way galaxy, and that only a quarter have heard that the universe is expanding. Such findings are lamentable, but of yet more serious concern is the fact that so few seem to understand science as a process.

To me, it doesn't matter so terribly much whether a student knows, say, how many planets there are in the solar system. For one thing, the astronomers are still arguing about whether Pluto ranks as a planet. For another, students can get the "right" answer in the wrong way. They may have learned from a book that the sun has nine planets, or heard a scientist say so on television in a booming authoritative voice. Scientific facts acquired in that fashion have no deeper meaning than a courtier's parroting of a king, or a professor's prattling that there can be no such thing as progress because Nietzsche and Schopenhauer said so. The scientific illiteracy decried in the newspapers—as when a camera crew, ambushing college graduates in their caps and gowns on graduation day, found that many didn't know what causes the seasons—is disturbing mainly as a symptom of a deeper problem, that they haven't learned how to investigate such questions. In the end, *what* you think is less important than *how* you think.

We're told that students aren't taught logic, that they don't know how to reason, but that's not the half of it. We had logic for thousands of years before we had real science, and what we learned was that logic can produce fantastic arrays of conclusions that have little to do with the world. There are, in other words, an infinite number of logically consistent universes; science asks in which of these universes we actually live. Students

who haven't learned this have failed to grasp science, regardless
of whether they can construct a syllogism or report that neon is
a noble gas. To them, science is a dangerous machine that oper-
ates in ways as mysterious as the casting of spells. Small wonder, then, that so many people fear and distrust science, so that
in movies and TV dramas—which prosper by mirroring the popular mind-set—scientists are more likely to die violently than
are members of any other profession, including gunfighting.

One motive in popularizing science is to help put people in
touch with their own evolving culture. This culture has, of
course, many other, older roots, such as art, religion, philosophy,
and history. These are more familiar, having served out
Whitehead's 1,000-year tenure, and so seem more natural. But
nothing is more natural than science, because nothing has done
more to show us how nature actually works. Part of the job of
the science popularizer is to help people realize this, so that
they may better live in an integrated world rather than in a
bifurcated world at war with itself.

We popularizers have not, however, done a very good job so
far. For decades now we've been making TV shows about science, and writing books and magazine articles and stories for
the science sections of the newspapers. Yet we still find ourselves living in a nation where, according to one study, fewer
than 7 percent of adults can be called scientifically literate by
the most generous definition, only 13 percent have at least a
minimum level of understanding of the process of science, and
40 percent believe in astrology.

Where did we go wrong?

Well, for one thing, we're in short supply. There are only
about 3,000 science writers in the United States, and perhaps
10,000 in the entire world, many of whom are freelancers who
moonlight at journalism while holding day jobs. The good news

is that there are more science writers today than there used to be, and that, to an increasing degree, we're being aided by scientists with a flair for writing who've also reached reasonably wide audiences. But we could use more recruits.

The pressing civic need for more science journalists is not, however, my main motive in urging young writers to consider covering science. Actually I don't much believe in practicing art for the public good; writing is hard enough to do for selfish reasons, without also burdening oneself with hopes for improving the lot of the general public. Lao Tzu said that governing a great nation is like cooking a small fish. Writing is, too. You don't have to clutter your mind with thoughts about how eating fish is good for people and that everybody should eat more fish. All you need to do is cook one fish, properly, at a time. I tell my students that science is a great story—the greatest story, in a sense—and that by writing about science they can open up vistas onto everything else. At least that's what happened to me.

When I started out as a journalist, among my colleagues were still to be found hard-boiled, old-style reporters who were proud of the fact that they lacked an education and who looked with scorn on copy that smacked of anything resembling thought. Some didn't even write: They got the story and then called it in to what was called the "rewrite desk," where it was turned into prose. Many of these men—and they were almost all men in those days—were excellent reporters, skilled and experienced in covering city hall, the courts, and the police beat. But science was not part of their world.

To give you a flavor of this era, let me quote a few lines from an interview conducted by a reporter for the *Wisconsin State Journal* in April 1929 with the visiting physicist Paul Dirac, who was famously taciturn and became even more so in response to these questions.

The reporter asked, "Now, Doctor, will you give me, in a few words, the lowdown on all your investigations?"

Dirac: "No."

Reporter: "Will it be all right if I it put this way: 'Professor Dirac solves all the problems of mathematical physics, but is unable to find a better way of figuring out Babe Ruth's batting average'?"

Dirac: "Yes."

Reporter: "What do you like best in America?"

Dirac: "Potatoes."[2]

We've gotten a bit more cerebral since then. Today most reporters have college degrees, and I'm happy to say that the majority of graduates in journalism in the United States today are women, so we journalists can at least boast that we're ahead of some of the sciences in tapping the brain power of the other half of the human species. In the broadcast news media, where the antiscience bias had been particularly strong, attitudes are beginning to change. And, in part, it's due to the Internet. On July 20, 1976, the *Viking* lander touched down on Mars at just the time when the network morning shows were on the air, live on the East Coast. Yet their producers declined to air the first live pictures from the surface of Mars. They said people weren't interested. Twenty years later, when *Pathfinder* landed on Mars, the number of people who visited the *Pathfinder* website to see its pictures amounted to more than the combined audiences of those three network morning shows.

So the producers were wrong. People *are* interested in science, and there *is* an audience for effective science reporting. One survey, conducted by the National Science Foundation in 1995, found that 86 percent of Americans agreed that "science and technology are making our lives healthier, easier, and more comfortable," and that 72 percent agreed that the benefits of science are greater than any harmful effects. In terms of confidence, the public ranked scientists second, behind only doctors, and followed by justices of the U.S. Supreme Court. (At the bottom of that list, by the way, were Congress, the press, and TV.)[3]

But, while there is an audience and it is being served somewhat better than in the past, the problem remains of how to get across what science actually is—how it works as a process, a way of approaching the world. It is to this issue that I wish to devote the remainder of this essay.

Many Americans want to know more about science, but are insufficiently acquainted with it to understand what scientific research is all about. For them, reading a report about stem cell research or balloon-borne mapping of the cosmic background radiation is like reading the score of a cricket match if you've neither played cricket nor watched it being played. To blame this situation on our schools is easy, and perhaps not entirely inappropriate. Only one-fifth of all American high school graduates have taken physics, and of those who did, only about a quarter were taught by someone who had a physics degree, and only a fraction of those teachers had ever done any research in physics. How well would high school football teams do if three-quarters of the coaches had never studied football and almost none had ever played the game?

But the schools simply reflect the wider reality, that science has not yet become integrated into the cultural mainstream. This makes it difficult to work science into television, movies, and the other media of popular culture, precisely because science is still novel, which means that the challenges posed to working scientists are unfamiliar to the general public. An audience filing into a theater to see a movie about love or sports or war brings to the experience some understanding of what it is to win and lose at love, sports, or war. Such is not the case with science.

How might this challenge be met?

Rather than preaching in terms of principles, I offer a few scenes written for film in which I tried to address this problem. It was a feature film, contemplated by the Walt Disney Studios, on the life of Albert Einstein. The film was never made; nor did I write the screenplay. But I did write a few scenes that were meant to convey some of Einstein's scientific ideas. Most involved special effects sequences. I envisioned the film as a musical without singing, in which the special effects took the place of the songs. In other words, they are intervals in which the audience is lifted out of the narrative flow of Einstein's life and transported to the realm of his thoughts.

I labored under few illusions then, and have few today, about the difficulty of this task. Nor do I represent this work as solving the problem of audience preparation that I've been describing. But it does represent an effort on my part to combine narrative and biography of scientific thought. So let me quote from a little of it.

This is the opening of the film:

Title on black. Fade title. Black screen. A sheet of blinding white light: We are seeing the big bang. Sound of thunder as the universe cools and expands. An epoch of darkness descends enlivened by gold, silver, and blue sparks of high-energy particles.

Theme music.

Galaxies begin forming from the darkness—dimly lit whirlpools with brilliant quasars flaring up at their centers. Shock waves sweep across the surrounding clouds of dust and gas, touching off the birth of billions of stars in each galaxy. Explosions rock the galaxies as they march apart from one

another in the ongoing expansion of the universe. In the background, the glow from the big bang slowly fades, stretching out
from blues to dull red by the expansion of the universe. The
glow dims to invisibility, leaving the blackness of space as a
backdrop to the galaxies.

We approach one galaxy—the Milky Way—as it rides through
space. Spiral arms sweep across the disk like brushfire. The
galaxy evolves from a predominance of brilliant blue stars to a
rich blend of blue, red, and yellow stars. We fly over the disk of
the Milky Way, over the golden central bulge, out across two
spiral arms, and dive into the disk, soaring through giant ruby-
colored clouds where new stars are being born under clouds
that tower up from the disk like coral heads.

Ahead lie the stars of the Sun's neighborhood. We pass through
the Hyades cluster in Taurus—the cluster we will see in the sky
in Einstein's childhood and later in the background of a solar
eclipse.

We approach the Sun, fall toward Earth, toward America, and
slowing, enter through the roof of Einstein's house on Mercer
Street in Princeton.

Extreme close-up: the yellowed face of an old compass.

Cut to interior, Einstein's study, night. The compass is held in
Einstein's hand. He spins it slowly, as enthralled as he was in
childhood at how the needle maintains its direction.

There ensues a series of scenes of the Einstein we all know—
the old Einstein, the legend, living out one day of his life in
Princeton. That night, after dinner, Einstein goes up to his
study. He's been forbidden to smoke, but he keeps a pipe and
tobacco hidden away; it was in searching for it the previous
night that he found the compass. He finds and fills his pipe,
lights it, then picks up the compass once more. This time we
start with the compass in extreme close-up, and we move in:

The needle fills the screen. It quivers with tension, responding to Earth's magnetic field. Closer: The wire wrapped around the compass needle fills the screen. Closer still: The metal working along the surface of the wire looks as pitted and gouged as the surface of the Moon. Metal crystals in the needle appear as we continue to move in; they look as regular as row house rooftops. A single crystal now fills the screen; it's like being inside a cathedral. We go down through layers to the level of the atoms, and we've reached the interior of an atomic nucleus. The nucleus decays, and the screen is suddenly flooded with white light, identical to the white light that we saw in the big bang at the beginning.

We reel backward through the zoom to again rest for the last time on the face of the compass.

Cut to interior, Einstein's boyhood house in the Munich suburbs, night, 1885. Einstein, age five, is in bed with a cold. The compass is in the hand of his father, Hermann. It's a present for the boy. Einstein turns it slowly in his hand, amazed that the needle always points north. Years later he would recall that his reaction was that "something deeply hidden had to be behind things."

EINSTEIN: How does it work?

HERMANN EINSTEIN: It responds to the Earth's magnetic field.

EINSTEIN: What's a magnetic field?

HERMANN EINSTEIN: It's a kind of energy wrapped around the world.

EINSTEIN: Around the whole world?

Cut to exterior, outside the Einstein home in the Munich suburbs, evening, a day or two later. Einstein, age five, stands outdoors on the grass watching the stars in the Hyades cluster as they come out in the darkening blue-black sky.

Close-up, Einstein. A shimmering grid of golden lines is reflected in his eyes. Offstage, his mother's voice calls him inside.

Einstein, from behind, ambling to the back door of the house while looking up at the stars. Above him, projected by his imagination, Earth's magnetic field stretches north to south across the sky. It's made up of glowing gold bands like wicker in a basket. As the boy goes in through the door, the golden grid doesn't quite fade away.

The next scene takes place a few days later, in the small dynamo plant that Einstein's father and uncle have set up in the backyard:

Interior, day, the electrical plant run by Einstein's father, Hermann, and his uncle, Jakob. The shop is noisy, busy, run on a shoestring, an adventure in the technology of the day, and one that in this case is destined to fail.

Hermann and Jakob are repairing a small dynamo. The young Einstein peers into it.

EINSTEIN: What does it do?

JAKOB: It makes electricity, Albert. Look in here. See these magnets? When we spin them like this, they generate electricity in
those coils of wire.

EINSTEIN: How does the electricity get from the magnets to the
wires, Uncle?

JAKOB: The spinning sets up an electromagnetic field.

EINSTEIN: Like the compass?

JAKOB (DELIGHTED): Why yes, that's right, Albert. The magnet
makes the electricity.

EINSTEIN: Is the electricity in there now?

JAKOB: No, not right now. The magnets have to be spinning for it

to make the electricity. That's why we're fixing it, so the magnets can spin.

EINSTEIN: So it's the motion that makes the electricity.

I'll skip ahead past a few scenes, including one in Italy when
Einstein, at age sixteen, imagines himself riding on a light beam
and hits on the central realization of what would become the
special theory of relativity. Let's go to his college years:

Interior, day, Herr Heinrich Weber's physics laboratory at the
Polytechnic Institute in Zurich. Light streams in through the
windows, but the lab is dark with shadows. Weber is demon-
strating a dynamo to the class. It's an important subject for
him, in part because his department at the institute was established with money provided by the builders of the big hydro-
electric dynamos that were then going up on principal rivers of
Germany. But Weber, like all other physicists of his day, fails to
fully understand how a dynamo works. This failure is particularly embarrassing because it prevents their accurately predicting the power output of the dynamo, dynamos into which so
much money is being invested.

The answer, Einstein is beginning to see, lies in extrapolating
out Maxwell's field equations. But Weber refuses even to teach
Maxwell's equations. In class, Einstein asks Weber questions
about the nature of the electromagnetic field in the dynamo,
referring to Maxwell's equations. Weber, infuriated, finally
orders Einstein out of the laboratory.

As Einstein, having gathered up his books, departs . . .

WEBER: The trouble with you, Einstein, is that nobody can tell
you anything.

Interior, of the corridor outside. Einstein runs down the hall.

Exterior, the Polytechnic Institute, day. Einstein bolts out the door, emerging from darkness into brilliant sunlight and leaps into the air. For the moment, he's free.

Cut to interior, the battery room of the Polytechnic Institute, late night. Herr Weber is conducting experiments in near-electrocution, using himself as the subject. He is seated in a sort of electric chair.

Weber attaches electrodes to his arms and legs, soaks the contacts in brine to ensure good conductivity, then gives the word and an assistant throws the switch. He is galvanized by a thousand volts of AC current. Einstein and friends, hidden in a warren near the ceiling of the lofty room, look on. The friends are amazed. Einstein is not. He's already learned that the world of human affairs is at least half-crazy.

Finally, two scenes having to do with general relativity:

Exterior, Principe Island off the African coast, day. Arthur Stanley Eddington and fellow astronomers have set up tents and a telescope and camera to record the eclipse. Gusting winds and intermittent rains hamper their efforts, flapping the tents. Scudding clouds obscure the Sun, now reduced to a crescent by the oncoming eclipse. At the last moment, a hole in the clouds opens up, permitting the astronomers to photograph the eclipse.

Exterior, the solar eclipse as it appears from space. The moon glides between Earth and the Sun, casting a giant shadow that rushes across the ocean toward the island. The stars of the Hyades cluster surround the eclipsed Sun, shining through its glowing pink-and-white corona. As the Sun moves in front of the cluster, the stars creep across the sky (as we've seen in a thought experiment earlier).

Exterior, the eclipse observation site. The total eclipse comes, an amazing sight. It has the assistants gaping. One calls out repeatedly to Eddington to look at it. Eddington, hunched over the camera, busy taking photographs, never looks up.

Cut to interior, a tent on the island, night. An improvised dark-room. Rain is falling outside. Wind billows the tent walls. Eddington is bent over the trays, developing the first photo-graphs of the eclipse. He takes a negative, still wet, sets it on the table, lays a transparent star map over it. The stars have shifted from their normal positions in the sky. He takes a sec-ond map, this of Einstein's prediction, lays it over the photo. The stars fall where predicted.

Cut to interior, Einstein's office in Zurich, September 27, 1919. Einstein is reading passages from a book on relativity to a stu-dent, Ilse Rosenthal Schneider.

EINSTEIN (READING ALOUD): "Einstein is completely unintelligible." What a wonderful admission by the author!

A messenger arrives with a telegram. Einstein opens it, reads it absently.

Close-up, the telegram. It reads, "Eddington found star dis-placement at rim of sun. Preliminary measurement between nine-tenths of a second and twice that value."

EINSTEIN (HANDING OVER THE TELEGRAM): This might interest you.

She reads it.

ILSE: How wonderful! This is just the result your theory pre-dicted!

EINSTEIN (TEASING HER): Did you doubt it?

ILSE: Well, no. Of course not, but what would you have said if the eclipse observation had not confirmed the theory?

EINSTEIN: I would have had to pity the dear Lord. The theory is correct.

What I'm trying to do here is, of course, to connect science to a large general audience, through a famous personality—*Time* magazine's "Man of the Century"—and to convey something of his thought through metaphor, storytelling, and special effects. One aim is to show that his thoughts are not just thoughts—that they're linked to observation and experiment. Einstein is unflappable, not because he's arrogant, but because he's brave. He's like a traditional hero embarking on a love affair or entering into combat. He knows that the verdict of experiment can be dangerous, as are love and war, but to face the danger in a cheerful, confident spirit is part of his job.

In conclusion, let me say that it's a pleasure to be able to write about someone like Einstein, and it's been my good fortune over the years to meet a number of leading scientists who are exemplary not only for their intellectual attainments but also for their character. Much of journalism has to do with unearthing secrets, getting people to tell you things they don't want known, picking through deceits and dissimilations to get at the truth. But, while scientists are human and hardly free from human weakness, science as a whole is much more open than are politics and finance. Science is a kind of white hole that gushes information, and the default position of most of the scientists I've met is to explain as much as possible, as clearly as possible, about what they think to be the facts about nature.

This was demonstrated to me with particular clarity one day years ago when we were making a film called *The Creation of the Universe*. We were at CERN, the European Organization for Nuclear Research, outside Geneva, Switzerland. The crew needed time to set up, leaving me with little to do but rehearse my lines, so I strolled down the halls, and every time I saw an open door with someone inside working, I went in and asked what they were doing. Not one of those scientists, interrupted in the middle of work by a total stranger, threw me out or tried to get rid of me with a brusque answer. Instead, each went to work telling me, as efficiently possible, all about his or her research. I wonder what the world would be like if more people were like that?

In addition, many scientists take time to write about science for general audiences. There are, of course, self-interested reasons for this, involving not just earning money or fame, and reporting to a public whose tax money helps pay for research, but also internal clarity. Niels Bohr insisted that physics, no matter how fancy, must ultimately be explained in ordinary language. Ernest Rutherford used to say that unless you can explain your theory to a barmaid, your theory is probably no good. One of the best of the scientist popularizers is the man we honored at Kipfest—not just for his intellectual achievements but also for his character. I've known Kip Thorne for many years, not only as a scientist—one of the world's supreme investigators of the expansive universe opened up by Einstein's relativity—but also as a man of integrity. In all that time, I've never once known Kip to say a discouraging word, to selfishly denigrate a colleague's work or unduly promote his own, to cut a corner to spare himself toil, or to obfuscate or distort or dissemble.

I do not myself believe that after death we shall find ourselves confronting St. Peter or some other judge who will call our lives into account, but the concept of such a judgment provides an excellent means of daily meditation. For in the end, the question is not just what you think and how you act and what you accomplish, but who had these thoughts and did these actions and accomplished these things. Diogenes, walking the agora in broad daylight with a lit lantern in hand and asked what he was doing, said, "Looking for a man." His point is still sufficiently obscure today that it's typically quoted as, "Looking for an *honest* man," but honesty only begins to cover it. Diogenes meant that he was looking for a *man*—someone who can say, "I know who it is who does my work and has my ideas, and he stands up for them, and takes responsibility for them." Had Diogenes encountered Kip in the agora that day, he could have extinguished his lantern.

Notes

1. Richard P. Feynman, *The Meaning of It All: Thoughts of a Citizen Scientist* (Addison-Wesley, Reading, Mass., 1998), p 49.
2. *Wisconsin State Journal,* April 30, 1929; quoted in Helge S. Dragh, *Dirac: A Scientific Biography* (Cambridge University Press, Cambridge, 1990), p. 73.
3. National Science Foundation, *Science & Engineering Indicators 1996;* quoted in *Skeptical Inquirer,* November/December 1996, pp. 6–7.

THE PHYSICIST AS NOVELIST

Alan Lightman

Kip Thorne has written some forty articles for the general public. In 1971, when I had the good fortune to become one of Kip's graduate students, I saw in the secretary's office one day a stack of reprints of his article "Death of a Star," which had won an award for popular science writing. Today, a great many scientists write for the public, but in 1972 the number was tiny. "Very interesting," I thought to myself. "So Kip is giving up some of his precious research time to write for the public." I took note. I revised my picture of the brilliant young physicist with the red beard who wore African-like tunics, who seemed to work day and night scribbling equations on his sheets of unlined white paper, who told us awe-struck students on the first day of class that we should call him "Kip," and who entered into scientific collusions with people named Yakov Borisovich Zel'dovich and Vladimir Braginsky and Igor Novikov. So Kip was a writer too. I took note. My impression of the writerly side of Kip was strengthened when fellow graduate student David Lee and I got back from him the draft of our first scientific paper, drowning in red ink. An attached note said: "How your papers are accepted, and the impact they have, will depend heavily on how they are written." I would venture to guess that few scientists, then or now, comment so carefully and helpfully on the quality of their students' writing.

At the time I was a graduate student in Kip's relativity group, I already had a strong interest in writing myself. In fact, since childhood, I had suffered double passions in the sciences and in the arts. In high school, I built rockets and I also wrote poetry. These dual interests tended to segregate my friends into two groups, and I often felt segregated even within myself. About seven or eight years after my Ph.D. in theoretical physics, I came out of the closet with my literary interests and began writing popular essays on science. The essay is an extremely flexible form of writing. In an essay, you can be informative or philosophical or personal or poetic. Soon I began to experiment with the essay, stretching its limits, and wrote some odd pieces that might be called fables—half fact, half fiction, still dealing with science but in an oblique way. Science as metaphor. Science as a way to view the world. About a decade ago, I left solid ground altogether and drifted into full-blown fiction.

One morning, I woke up and discovered that I had become a member of a second community. I use the word "community" here loosely because the community of writers is very different from the community of scientists. An active scientist stays in close touch with dozens of other scientists, often does research in a university or laboratory setting among other scientists, gives seminars on new research, telephones or e-mails professional colleagues on a daily basis, exchanges papers before publication, and goes to several conferences a year. By contrast, writers work in isolation. The two major organizations for writers, PEN and the Author's Guild, exist mainly for promoting the legal and political rights of writers and for giving prizes. Most writers do not attend meetings. Most writers write at home, alone.

A novelist may spend five years working on one book. During those five years, he might talk to one or two other writers every six months, talk to his agent a couple of times, talk to an editor in the fourth year. Once in a while, he might attend a book festival or give a reading with a couple of other writers. A novelist lives in the desert. He has evidence of the existence of other novelists mainly through the occasional footprints he stumbles upon, in the form of books and reviews. He reads other writers' books with admiration and jealousy, then goes back to his one-man tent. That's the community of writers.

As a member of these two communities, such as they are, I've been fascinated by their different ways of working, their different ways of thinking, their different approaches to truth. And, at the same time, by the similarities. Boston, where I live, is a town that has lots of both kinds—writers and scientists—and sometimes, when I'm riding the subway, I play a game with myself and try to pick out the scientific or literary types just by their looks. That fellow over there staring out the window into the blackness with a puzzled expression on his face, wearing the green checkered pants and the plaid nylon shirt, the pen protector with four pens in his breast pocket, the beaten-up briefcase that should have been given to Good Will a decade ago—I'll bet he's a theoretical physicist. And that guy in the corduroys and tweed jacket with tousled hair and a carefully tended two-day-old beard, his body stretched out in a long sulk, a notebook in his hand, scribbling as he keenly peruses each passenger on the car, looking over at me now as I look over at him—I'll bet he's a writer. . . . But I find that these stereotypes in appearance don't always work. I put away my own notebook and estimate how many minutes until the Kendall Square stop.

———————

A big distinction I've found between physicists and novelists, or scientists and artists generally, is in what I'll call the "naming of things." Roughly speaking, the scientist tries to name things and the artist tries to avoid naming things. There are many facets to this distinction. I'll describe a few.

To name a thing, you've gathered it, you've distilled and purified it, you've attempted to identify it with clarity and precision. You've put a box around it and claimed, "What's in this box is the thing and what's not is not." For example, consider the word "electron," a type of subatomic particle. As far as we know, all of the zillions upon zillions of electrons in the universe are identical. There is only a single kind of electron. And to a modern physicist, the word "electron" means a particular equation, the Dirac equation with field operators. That equation summarizes, in precise mathematical and quantitative form, everything we know about electrons, every interaction, every blip that will be measured by our atom smashers and gauges and magnetometers. The energies of electrons in atoms of various types, the precise deflections and twists of electrons by particular magnetic and electric fields, the tiny effects of electrons and their antiparticles materializing out of nothing and then disappearing again—all of that can be predicted accurately to many decimal places by the Dirac equation with field operators. You can discuss this or that aspect of electrons, whether an electron spins like a top or turns inside itself, whether it orbits or hovers, whether it spreads out like a wave or concentrates itself like a poppy seed, but the Dirac equation has a much more precise and objective representation of the electron. In a real sense, the name electron refers to that equation. Modern physicists know and love the Dirac equation. Every physical object in the universe scientists aim to express with as much precision. It is a great comfort, a feeling of power, a sense of control, to be able to name things in this way.

The objects and concepts the novelist deals with cannot be named. The novelist might use the word "love" or "fear," but those names don't summarize or convey much to the reader. For one thing, there are a thousand different kinds of love: There's the love you feel for a mother who writes you every day during your first month away from home in summer camp. There's the love you feel for a mother who slaps you when you stumble into the house drunk after driving home from the prom and then embraces you. There's the love you feel for a man or a woman you've just made love to. There's the love you feel for a friend who calls to give you support after you've just split up with your spouse. And on and on. But it's not just the different kinds of love that prevent the novelist from truly naming the thing. It's that the sensation of love, the particular sensation out of the thousands of different kinds of love, the particular ache, must be shown to the reader, not named, but shown through the actions of the characters.

And if love is shown rather than named, each reader will experience it, and will experience it in her own individual way. Each reader will draw on her own adventures and misadventures with love. Love means one thing to one person and a different thing to another. Every electron is identical, but every love is different. The novelist doesn't want to eliminate these differences, doesn't want to clarify and distill the meaning of love so that there is only a single meaning, like the Dirac equation, because no such distillation could represent love. And any attempt at such a distillation would destroy the authenticity of the reactions of readers, would destroy that participatory creative experience that happens when a good reader reads a good book. In a sense, a novel is not completed until it is read by a reader. And each reader completes the novel is a different way.

Now, there is more to this business of naming and not naming than the sameness of electrons versus the varieties of love. Even a single reader changes from one moment of his life to the next. His experiences and relationship with the world changes, and thus the meaning of a story or a character or even a single word changes for him over time. I once went to a conference of the Modern Language Association, which is the top professional organization for literary critics and English professors, and there was a session on science as literature. One of the professors stood up and said that the ideal scientific text would be one that is so concise and clear and exact—dealing with a world of exactness as it does—that a reader would have to read it only once. But the ideal literary text, such as a novel, would be one that a reader would need to read over and over, because it would have the complexities and ambiguities of human behavior and, with each new reading, the reader would be at a different point of his life and appreciate different things and get something new out of the book.

I'll give another illustration of the difference between naming and not naming. Let me represent science by expository writing. Like science, a piece of expository writing takes a reductionist and reasoned approach to the world. You have a position or argument, you structure your argument in logical steps, you amass facts and evidence to convince your reader of each assertion, and you lead your reader in a more or less direct route from some starting point to a finish of increased comprehension. We all learn that in expository writing it is excellent form to begin each paragraph with a topic sentence. A topic sentence, in effect, names the idea of the paragraph. Begin by telling your reader what she is going to learn in the paragraph and how to organize her thoughts.

But in fiction writing, a topic sentence is usually fatal. Because the power of fiction writing is emotional and sensual.

You want your reader to feel what you're saying, to smell it and hear it, to be part of the scene. You want your reader to be blind-sided, to let go and be carried off to that magical place. Every reader will travel differently, depending on his own life experiences. Telling your reader at the beginning how she's supposed to think about something cancels the trip. And if there is an idea involved—and many novels do deal with ideas as well as with characters and narrative—you want not to state your idea baldly but to let it seep in slowly and gradually, around the edges, so that your reader must sweep through the terrain over and over again searching for meaning, haunted. With a topic sentence, you don't leave room for your reader's own imagination and creativity. The difference between these two kinds of writing can be stated in terms of the body. In expository writing, you want to get to your reader's brain. In creative writing, you want to bypass the brain and go for the stomach or heart.

A pattern of thinking closely related to naming is the tradition of framing problems in terms of questions and answers. Scientists usually work by finding interesting problems and then breaking up those problems into pieces that can each be stated in terms of a definite question with a definite answer. In fact, much of the game of science is to pose a problem with enough precision and clarity to guarantee a solution. The world is then built, piece by piece, from these solvable problems. For example, a typical scientific problem might be: how does a star change in time? One piece of this problem would be: what is the structure of a star with a given chemical composition and given pressure and density at its center? This is a well-posed problem with a definite solution. Another piece of this problem would be: what is the rate of nuclear reactions of a given mixture of hydrogen and helium gas at a given temperature and density? And so on. Scientists are taught from an early point in their apprenticeship not to waste time on questions that do not have clear and definite answers.

But artists often don't care what the answer is because definite answers don't exist. Ideas in a novel or painting are complicated with the intrinsic ambiguity of human nature. Indeed, the exquisite contradictions and uncertainties of the human heart make life interesting. They are why the actions of characters in a good novel can be debated endlessly, why we react to Gore or Bush on a gut level, why God held the apple in front of Eve and then forbade her to eat it. For artists, there are many interesting questions without answers, such as What is love? or Would we be happier if we lived to be 1,000 years old? or Why does a sunset appear beautiful? In fact, for many artists, the question is much more important than the answer. As the poet Rainer Maria Rilke wrote a century ago, "We should try to love the questions themselves, like locked rooms and like books that are written in a very foreign tongue."[1]

One consequence of this difference in having definite answers or not shows up in the workday of the scientist versus that of the artist. When I was active as a physicist, I was sometimes grabbed by a science problem so that I could think of nothing else, consumed by it during the day and then through the night, hunched over the kitchen table with my pencil and pad of white paper while the dark world slept. I was tireless, electrified, working on until daylight and beyond.

As a writer, even when I am writing well, I cannot write more than six hours at a time. After that I am exhausted, and my vision has become clouded by the inherent subtleties and uncertainties of the work. Then I must wait for the words to shift and settle on the page and my own strength to return.

But as a scientist, I could be gripped for days at a time, days without stopping, because I wanted to know the answer. I wanted to know the telltale behavior of matter spiraling into a black hole, or the maximum temperature of a gas of electrons and positrons, or what was left after a cluster of stars had slowly lost mass and drawn in on itself and collapsed. When in the throes of a new problem, I was compelled because I knew there was a definite answer. I knew that the equations inexorably led to an answer, an answer that had never been known before, an answer waiting for me. That certainty and power, and the intensity of effort it causes, cannot be found in most other professions.

As a person trained in the sciences and in the various ways of naming things, I've had a constant struggle as a fiction writer. The great push and pull in my writing life, and in my life as a whole, has been the tension between the rational and the intuitive, logic versus illogic, certainty versus uncertainty, linear versus nonlinear, deliberate versus spontaneous, predictable versus unpredictable. I experience this tension as a constant twisting of my stomach when I'm aware of my body and always as a mental commotion. I've learned to live with the discomfort. It may actually be a source of strength. Over time, I have come to believe that both certainty and uncertainty are necessary in the world. Perhaps this idea is obvious to most people, but it's not so easy to recognize for someone trained in the sciences.

Even as a writer, there's a big difference between nonfiction and fiction. When I write essays or reviews or articles about science, I know that I can research a subject, collect my material, outline a presentation. In short, I feel in control. I know pretty much where I'm going. When I write fiction, I do not feel in control. I cannot predict what will happen. I know that I must give my fictional characters enough freedom and life that they can surprise me. After that, a character may decide that she doesn't like my plot. She may do something that throws a whole scene into disarray, maybe the whole book. Thank you for that, I say, and wince silently. Fiction writing makes me nervous. It makes me happy, but nervous.

Here's how a person torn between the virtues of certainty and uncertainty creates a character. In my first draft, I have the outlines of a character. But only the outlines, because a character really defines herself by the way she acts in various situations, and if I don't know exactly how a character is going to act in advance, I don't know the character. After the first draft and all of its unpleasant surprises, I have a better idea of the character. In the second draft, with my deeper knowledge, I revise the character, improving dialogue that no longer is quite right, altering actions that now seem inconsistent. After the second draft, I have an even deeper understanding of the character and repeat the process. In this way, the character is built up by a series of approximations.

As I now think about this method for developing a character, I'm suspicious. It's too logical. Creating good characters is not my strong point. I find it easier to create scenes and atmosphere. Originality I value above everything else. As might be expected from someone trained in the sciences, ideas play a major role in my writing, but ideas, in fiction, must be handled like high explosives. Ideas can destroy a short story or novel when the characters become mouthpieces for the intellectual grandstanding of the novelist. It is better that a novelist's intellectual intentions not barge through the front entrance, but slip in quietly through a back door.

I want to say something now about the substantial common ground of the physicist and the novelist.

The folklore is that novelists make up everything and physicists make up nothing. Both views are false. Creative imagination and inventiveness have always been hallmarks of good physicists, just as of good novelists. On the other hand, novelists must conform to a certain body of recognized truth about human nature, just as physicists must adhere to truth about nonhuman nature.

Theoretical physicists, especially, work in a world of the mind. It is an abstract, mathematical world. Physical reality is represented by simple models that can be visualized, or mathematical equations that can be written down on a piece of paper. For example, a physicist can imagine a weight hung from a spring, bouncing up and down, and can fix this mental image with an equation. If friction with air becomes an unwanted nuisance, just imagine the weight in a vacuum. No real weight on a spring exists in a perfect vacuum, but thousands do in the minds of physicists.

Einstein often emphasized the importance of what he called the "free invention" of the mind. The great physicist believed that we cannot arrive at the truths of nature only by observation and experiment. Rather we need to create concepts, theories, and postulates from our own imagination and only later confront these mental constructions with experience.

One of the best illustrations of Einstein's free invention in science was his work on the special theory of relativity—a theory that led to radical new concepts of time and space. That work begins with the stunning postulate that the measured speed of a light ray is always the same, independent of the motion of the emitter or the observer. Einstein called that statement a "postulate" because there was no experimental evidence that required it. In fact, most experimental evidence suggested the contrary. Either a moving thing is pitched, like a baseball, in which case its speed past an observer depends on the speed of the pitcher relative to the observer, or a moving thing travels as a wave, like a water wave, in which case its speed past an observer depends on the observer's own speed through the water.

Einstein's postulate about the constancy of the speed of light rays violated all common sense. Yet he realized that common sense could be in error when it came to extremely high speeds, like the speed of light, and he made an imaginative leap with his postulate. In deriving the consequences of his strange postulate, he found that the standard ideas about time—that time is absolute, that a second is a second is a second—had to be revised. Here again, experiments could not have provided any clue because the discrepancies in the ticking rate of clocks were too small to measure. To be sure, Einstein was influenced by certain experiments with electricity and magnetism and with the knowledge that light is a traveling wave of electromagnetic energy, but none of these experiments required his daring and creative postulate.

A more recent example of the use of invention in physics is string theory. Here, physicists have proposed that the fundamental units of nature are not subatomic particles, like electrons, but tiny one-dimensional strings. The typical length of one of these hypothesized primal strings is 10^{-33} centimeters, one-hundredth of a billionth of a billionth of the size of the nucleus of an atom. Needless to say, none of these incredibly small strings has ever been seen, nor is one likely to be. There's one other detail about strings. They inhabit a universe of at least nine dimensions, six beyond the usual three. We don't see the extra dimensions because they are curled up into ultratiny loops.

When Yoichiro Nambu, Holger Nielsen, Leonard Susskind, John Schwarz, and Joel Scherk first proposed the ideas of strings in the early 1970s, they were using a great deal of imagination. They were trying to understand the basic forces of nature. But no experimental facts required the postulate of strings versus particles, and certainly no observations had ever suggested that we live in a nine-dimensional world. Most people have enough trouble grappling with length, width, and breadth. These physicists were following Einstein's method of letting their minds spin freely, making postulates, and then working out the consequences of those postulates. To date, there is no experiment that we know of capable of truly testing string theory. In fact, the theory hasn't even made any definite predictions. Yet some of the best theoretical physicists—artists all of them—are working on string theory, creating and inventing out of their heads.

Of course, physicists can't make everything up, even when they're inventing new theories. There is already a huge body of facts known about the physical universe, and these facts may not be contradicted. Richard Feynman put it well in his little book *The Character of Physical Law:* "What we need is imagination, but imagination in a terrible straight jacket. We have to find a new view of the world that has to agree with everything that is known, but disagree in its predictions somewhere. . . ."[2]

Just as the physicist has to agree with some known facts when he or she is making up new things, so does the novelist. But what is the straightjacket of the novelist? It is the large catalog of known behavior and psychology of *Homo sapiens*, a catalog we sometimes call human nature. These are the facts of emotional truth that the novelist is bound by.

Let me give an example. Suppose the novelist has created a character about forty years old, married with two children, a man who has just attended a Christmas party with his wife. This fellow—we'll call him Gabriel—is not completely sure of himself. When he first arrives at the party, he worries that he has accidentally insulted the housekeeper's daughter. Then he worries about how his after-dinner speech will be received. After the party, he and his wife walk to a hotel where they are staying for the night. They've left their children with a cousin in a neighboring town. It's snowing. Gabriel's wife, Greta, has been rather quiet during the evening. But Gabriel, walking alone with her, is overcome with admiration, love, and desire for her. He looks at her tenderly and remembers the precious moments of their life together. He wants to remind her of them, to make her forget the many years of their dull existence together, the routines of daily life, the children, her household cares. They walk up to the snowy hotel late at night, and climb the stairs to their room, lit only by candlelight.

Gabriel is now burning with desire for her. He wants her to offer herself to him with an equal desire, but instead she turns away from him and begins to cry. He asks her what's wrong, and eventually she says that a sad song at the Christmas party reminded her of a young man she once knew long ago in her youth. Gabriel begins to feel a vague dread in his stomach but continues asking his wife questions about this young man of the past. He was seventeen, Greta says, he worked in the gasworks, and he was a tender, delicate young man with big, brown eyes. They used to go walking together, in the country. Gabriel asks Greta if she was in love with this boy, and she replies that she was "great with him at that time." Then she says that he died at age 17. What did he die of, so young? Gabriel asks his wife. "I think he died for me," answers Greta. She stops talking, is overcome with grief, and flings herself down on the bed, sobbing.

This scene I've described is, in fact, the last scene of James Joyce's famous story "The Dead." How will Joyce end the scene? What will be Gabriel's reaction to his wife's confession? Suppose Gabriel shows no reaction. Would we, as readers and with our own life experience, believe this? No. This ending would be false. Or suppose Gabriel feels superior to Greta's dead lover of the distant past and dismisses her pain. This reaction, too, would be false. Or suppose Gabriel becomes furious with his wife, takes her confession as if it were adultery, and decides to leave her. This is a possible ending, but it doesn't square with what we already know of Gabriel. The ending that Joyce actually writes is this: Gabriel realizes that his wife has always loved this long-dead boy more than she ever loved him, realizes what a poor part he, her husband, has played in her life against this memory, realizes that he himself has never loved any woman with the force that his wife has just shown. Gabriel can only sag against the glass window, listening to the breathing of his wife as she sleeps, watching her as if he and she have never been man and wife. We believe this ending, we know that it is true, even in fiction, because it accords with our knowledge of human nature, our personal experience with life. And it causes us anguish.

Both the novelist and the physicist are seeking truth—for the novelist, truth in the world of the mind and the heart; for the physicist, truth in the world of force and mass. In seeking truth, both the novelist and the physicist invent. Both kinds of invention are important; both ultimately must be tested against experiment. The tests in physics are more objective and final. No matter how beautiful a physicist's invention, it suffers a terrible vulnerability: it can be proved wrong. This terrible vulnerability to experiment is why I cannot agree with the school of the philosophy of science that says all of science is a human construction. Scientists often wish powerfully for some theory to be true that is later proved wrong by the facts. Aristotle's idea that the planets move in perfect circles was simple and elegant, but proved wrong by Brahe, Kepler, and Newton.

A novelist's story or characters cannot be proved wrong, but they can ring false and thus lose their power. In this way, the novelist is constantly testing her fiction against the accumulated life experience of her readers.

An experience that the physicist and novelist share, a most extraordinary experience, is the creative moment.

We all know that a great deal of the activity of scientists and artists is not especially creative: working out the details of a calculation, checking the lubrication of a seal on a vacuum pump, researching the locale for a novel, laying in a background tint. But there are other periods, which might last only a few seconds or perhaps hours, when something different happens, when the scientist or artist is in the grip of inspiration—and here I think the experience is quite similar

I write in two places. One is an island in Maine. From my writing desk, I can see the ocean fifty feet away, I can see ospreys and bayberry bushes and the pine-needle path leading from my house down the hill to the dock on the water. The other place where I write is a storage room off the garage of my house in Massachusetts, a room the size of a large closet, damp, closed off, and without windows. There I can see nothing except the white cement wall one foot from my desk. Both places have served me equally well in my writing because after twenty minutes at work, I have vanished and reappeared far away in the imaginary world I've created, oblivious to my former surroundings. In this magic trick of transportation, I become oblivious not only to my actual environment but also to my self, my ego, my body.

What a strange and beautiful paradox of creativity, that we dive deep into ourselves to create something, drawing on what is most private and personal, and completely lose ourselves in the process. When I am writing, I forget where I am and who I am. I become a pure spirit; I melt into all the other spirits who have ever created. These moments, I think, are the closest a human being can come to immortality. These moments are when I am happiest.

My first experience with the creative moment in science occurred during my years as a physics graduate student here, at the California Institute of Technology. I was twenty-two years old. When you're getting a Ph.D. in science, in addition to taking courses, you must solve an original research problem important enough to get published. One of my first research problems as a graduate student had to do with the behavior of gravity, whether gravity must be equivalent to a warping in the geometry of space and time.

After an initial period of work and study, I had succeeded in writing down all of the equations to be solved. But then I hit a wall. I knew I'd made a mistake, because a result at the halfway point was not coming out as it should, but I could not find my error. And I could not go on. Day after day, I checked each equation, pacing back and forth in my little windowless office, but I didn't know what I was doing wrong, what I had missed. This confusion and failure went on for months. Unlike all the other problems I had encountered in school, I could not look up the answer in a book. The answer to this problem was not known. I was obsessed by my research problem; I brooded about it day and night. Some days I didn't leave the office. I ate lunch and dinner there. I kept cans of tuna fish in the drawers. I stopped visiting my friends. I was beginning to doubt my abilities. I was beginning to believe that I didn't have what it took to be a scientist.

Then one morning—I remember it was a Sunday morning—I woke up about 5 A.M. and couldn't sleep. I was in my apartment, not my office. I felt extremely excited. Something was happening in my mind. I was thinking about my science problem, and I was seeing deeply into it. The physical sensation was that my head was lifting off my shoulders. I felt weightless. I was floating. And I had absolutely no sense of myself. It was an experience completely without ego or any thought about consequences or approval or glory. I had none of those feelings. I did have a feeling of certainty. I had a strong sensation of seeing deeply into this problem and understanding it and knowing that

I was right. That's an amazing aspect of the creative moment—
this inner knowing that you're right, this compelling sense of
rightness.

So with these sensations surging through me, I tiptoed out of
my bedroom, almost reverently, afraid to disturb whatever
strange magic was going on in my head, and went to the
kitchen. There was a table there, and I got out the pages of my
calculations. Just a tiny bit of daylight was starting to come
through the window. Although I was oblivious to everything
around me, the fact is that I was completely alone. I don't think
any other person in the world would have been able to help me
at that moment. And I didn't want any help. I had all of these
sensations and revelations going on in my head, and being alone
with all that was an essential part of it. I knew things that no
one knew. And this knowledge made me feel powerful, like I
could do anything. I was in this fantastic condition of seeing.
Since I had no sense of myself, there was no "I" doing the seeing,
no see-er. It was just pure seeing.

I sat down at the table and began working, making simplifica-
tions here and there that I understood were good approximations
because I could see the whole thing. Somehow, perhaps for
weeks, my unconscious mind had been taking secret paths, trying out different possibilities and connections, and was now
spilling itself. After a while at the kitchen table, I had solved my
research problem. I went out of the room, feeling stunned and
powerful. Suddenly I heard a noise and looked up at a clock on
the wall and saw that it was two o'clock in the afternoon.

I mentioned the inner feeling of certainty in the creative
moment. I've experienced it both as a physicist and a novelist,
and I think that sense of certainty is connected to the power of
beauty in the human psyche. Physicists are driven by aesthetics
as much as novelists are. When Einstein was searching for a uni-

fied theory, combining gravity with electromagnetism, he wrote to his friend Paul Ehrenfest saying: "The latest results are so beautiful that I have every confidence in having found the natural field equations of such a variety."[3] And Feynman, who was not given to sentimentality, said in *The Character of Physical Law* that one of the important things in guessing new laws of physics is to "know when you are right. It is possible to know when you are right way ahead of checking all the consequences. You can recognize truth by its beauty and simplicity."

The physicists and novelists I've known have at least one more thing in common: they do what they do because they love it, and because they cannot imagine doing anything else. This compulsion is both blessing and burden. A blessing because the creative life is filled with beauty and not given to everyone, a burden because the call is unrelenting and can drown out the rest of life. This mixed blessing and burden must be the "sweet hell" that Walt Whitman referred to when he realized at a young age that he was destined to be a poet. "Never more shall I escape."[4] This mixed blessing and burden must be why Chandrasekhar continued working on physics until his mideighties, why Hans Bethe still calculates on supernovae at age ninety, why a visitor to Einstein's apartment in Bern found the young physicist rocking his infant with one hand while doing mathematical calculations with the other.

When a beginning poet wrote to Rilke and asked whether he should continue to write, Rilke answered that he should write only if he could not not write: "Search for the reason that bids you to write; find out whether it is spreading out its roots in the deepest places of your heart, acknowledge to yourself whether you would have to die if it were denied you to write. This above all— ask yourself in the stillest hour of your night: must I write?"[5]

I remember a trip to Kip's cabin on Palomar Mountain almost thirty years ago. It was a hot summer day. We had a lot to do when we arrived—hiking up a hill, hauling in boxes of food and beer for the weekend, Kip's graduate students and postdocs fussing with their tents and sleeping bags and bug repellent. As I recall, people took turns swinging from a rope or a swing attached to a tall tree. A fire was made for cooking. Someone had a grill, charcoal, matches, chicken, and steaks. I looked around for Kip and finally found him off by himself, sitting quietly in a foldable chair by a big rock. He was hunched over a white pad of paper, scribbling equations, oblivious to the world, happy, doing what he loved more than anything, doing what he must do, blessed and burdened at the same time. It was another good lesson for a young student.

Notes

1. Rainer Maria Rilke, *Letters to a Young Poet* (W. W. Norton, New York, 1962), p. 35.
2. Richard Feynman, *The Character of Physical Law* (MIT Press, Cambridge, Mass., 1965), p. 171.
3. Albert Einstein, letter to Paul Ehrenfest, September 24, 1929; quoted in Albrecht Fölsing, *Albert Einstein: A Biography*, trans. Ewald Osers (Viking Books, New York, 1997), p. 606.
4. Walt Whitman, "Out of the Cradle Endlessly Rocking," from the *Sea-Drift* section of *Leaves of Grass* (published by the author, 1855; many subsequent editions and reprints).
5. Rilke, pp. 18–19.

GLOSSARY

This glossary gives information about some of the technical terms that appear in this book, along with some guidance in locating them in the essays. Since the technical scope of the essays is fairly narrow, the number of terms dealt with here is fairly small, but the explanations go beyond definitions and can stand by themselves. In that sense this glossary is complementary to the one with a large number of short entries that appears near the end of Kip Thorne's *Black Holes and Time Warps: Einstein's Outrageous Legacy* (W. W. Norton, New York, 1994).

Words in boldface type are references to entries defined elsewhere in the Glossary.

black hole horizon

A "horizon," or "event horizon," is a closed surface that divides spacetime into an exterior region and an interior region called a black hole. The property that defines a horizon is that no signal or influence from the interior can reach the exterior. For a more complete discussion, see the section "Horizons and Black Holes" in the Introduction.

blueshift

The terms "redshift" and "blueshift" describe the difference between the frequency of light emitted by some source, such as a star, and the frequency received by some observer, such as an astronomer. Equivalently, the terms refer to the difference between the energy with which photons are emitted and the energy with which they are received.

If a star is moving away from Earth, then Earth-based astronomers will receive redshifted light, light that has lower frequency, or is redder, than the light produced by processes in the star. A photon received will have an energy lower than that with which it was emitted. The opposite happens if the star is moving toward Earth: the light received on Earth will be shifted to higher, bluer frequencies, and photons will be more energetic when received than when emitted.

Redshifts and blueshifts are also caused by gravitational fields. If we stand at the bottom of a tall lighthouse, the photons that reach us will gain energy as they "fall" toward us. They will be more energetic and hence bluer than when they are produced at the top of the lighthouse.

In his essay, Stephen Hawking points out that light will become blueshifted on each cycle of a loop in which it returns to an earlier event.

Cauchy horizon

Central to the ideas of relativity is causality, the way in which events can influence each other. An "event" is a "point" in spacetime, that is, a location in space at a particular time. One event can influence a second event if, in principle, a signal could be sent from the first to the second event at or less than the speed of light.

Suppose we take all points in spacetime at some instant of time; that is, we take a **surface of constant time.** In mathematical relativity, those spacetime points that are influenced by our surface of constant time are called the "Cauchy development" of that surface of constant time (named for the French mathematician Augustin-Louis Cauchy, 1789–1857). In a sense, these points are the answer to the question, What will develop from the information on this surface of constant time?

The normal expectation is that *all* points to the future of this surface will be in its Cauchy development. But, as Stephen Hawking points out in his essay, there are spacetimes in which this is not the case; in these spacetimes there are surfaces of constant time that do not determine all regions to the future. For such spacetimes Stephen Hawking has introduced the term "Cauchy horizon" to mean the boundary of the regions that can be determined. As he points out, Cauchy horizons occur in some black hole solutions of Einstein's equations. He also shows that under some circumstances, a Cauchy horizon is inevitable if spacetime is to contain a region with **closed timelike curves.**

A Cauchy horizon differs from the horizon (full name "event horizon") of a black hole. (See the section "Horizons and Black Holes" in the Introduction.) The two kinds of horizons, though, share the property of separating spacetime into distinct regions.

closed timelike curves

All of the events on a worldline have a timelike connection to one another (see **timelike and spacelike**) so the worldline of an object can be called a timelike curve through spacetime. By certain methods, such as going through a **wormhole,** the object can return to an event a second time. This means that the object's worldline is closed; that is, it forms a loop.

The paradox of a closed timelike curve is that each event on it is in both the past and the future of every other event on the curve. The resolution of this paradox is that a wormhole allows for nontrivial connections of events in spacetime. These connections cause us no difficulty if the curve is spacelike, like Earth's equator. On such a path you can walk constantly in one direction, say east, and yet return to your starting point. This is possible because the connections between points on Earth's surface are those of a nontrivial geometry, that of a spherical surface. That can happen to a timelike curve also if the geometry is connected up the right way.

cosmic censorship

The development of a **singularity** in spacetime causes a terrible problem for gravitation theory. Scientific principles can govern how things change in time only if a complete set of principles is given, enough "laws" to specify the behavior of spacetime and everything in it. But by its very nature of containing infinite quantities, such as **energy density** or curvature, a singularity cannot be described by such principles. If a singularity forms in spacetime, then physical law will lose the power of predicting what comes next. (This failure of predictability can lead to a **Cauchy horizon.**)

Black holes offer a way to make this breakdown tolerable. Nothing inside the event horizon of a black hole can influence anything outside the horizon. A singularity inside the horizon would therefore not cause problems in predictability outside the horizon. In (almost) all examples known, singularities form inside horizons, and the horizons "shield" the exterior regions from unpredictability. Singularities that are not shielded in this way are called "naked." Cosmic censorship is the idea (or hope) that no "naked singularities" can form in realistic situations, that the laws of physics will censor a singularity by hiding it behind a horizon.

cosmic string

A cosmic string is a theoretically proposed filament of matter and energy having zero cross section. Since matter and energy curve spacetime, a cosmic string influences the spacetime around it. A cosmic string can be considered simply a thread of extremely **exotic material** confined so tightly that it takes up no volume.

The reason that cosmic strings are interesting, however, lies in another viewpoint. The way in which spacetime is distorted around a cosmic string corresponds to the string being a minor "defect" in spacetime. An analogy for this kind of defect can be constructed with paper and scissors. Cut a wedge out of a disk of paper as if you were taking a single serving from a pie. Next, smoothly glue together the cut edges of the remainder of the disk to form a cone. The defect here is the tip of the cone; it is a zero-dimensional defect in the two-dimensional paper. Due to that defect, the cone is different in some ways from a flat piece of paper. In somewhat the same way, a cosmic string is a one-dimensional defect in four-dimensional spacetime. In his essay, Stephen Hawking shows that cosmic strings, rather than **wormholes,** can be used to create **closed timelike curves.**

Cosmic strings have no direct connection with "string theory," a theoretical approach to understanding the details of fundamental particles and forces. Cosmic strings can be as long as the universe is large. String-theory strings are smaller than the smallest elementary particle.

cosmological constant

Einstein invented a modification in the way **general relativity** relates the curvature of spacetime to the amount of matter and energy that are present. A quantity called the "cosmological constant" governs the amount of this modification. If this constant is set to zero, the result is Einstein's original theory of relativity (called his "special theory of relativity"). The form of the modification is the same as it would be if the universe were uniformly filled with a very low density of material with unusual properties. Because of this, the cosmological constant is sometimes considered to be related to the physical properties of so-called empty space. It is as if empty space (the vacuum) had an intrinsic energy.

At present, it is difficult to make astronomical observations compatible with Einstein's standard theory, and the cosmological constant is receiving much attention from scientists. The observations are also driving physicists to consider new kinds of "dark energy" that might be filling the universe.

In his essay, Stephen Hawking refers to the strange properties of the Gödel spacetime, a solution of Einstein's theory with a cosmological constant.

curved and flat

In a two-dimensional plane like a blackboard, you could in principle construct a set of straight parallel lines with unchanging distances among them. You could then construct a second set with the same properties and have all the lines of this second set perpendicular to the lines of the first set. Because you can do this, the geometry of the plane is said to be "flat." This construction cannot always be carried out, even in principle. On the surface of Earth, for example, it cannot be done. A geometry for which it cannot be done, (that is, one that is not flat) is called "curved." In a three-dimensional geometry, we need to ask whether it is possible to construct three sets of lines, each set perpendicular to the other two sets. In each set, the lines must remain parallel as they are extended, no matter how far. This concept of flat and curved can be extended in this way to any number of dimensions, and applies to spacetime as well as to space.

cutoff

See **two-point function.**

energy density, energy-momentum tensor

According to **general relativity,** Einstein's theory of gravitation, the curvature of spacetime, and hence the force of gravitation, is generated by the matter and fields (such as electric fields) that are present. How sharply spacetime is curved is determined by how tightly matter and fields are packed into spacetime. Of primary importance is the density of energy in spacetime, the amount of energy per unit volume.

For ordinary matter and fields, the energy density is zero or positive, and the common wisdom is that negative energy densities are impossible in a practical sense. This is unfortunate for time travel, since **wormholes** require a negative energy density, but there is some hope for wormholes. According to **quantum theory,** the quantum fluctuations of fields, in some circumstances, allow for the possibility of negative energy densities. Much of the recent debate about time travel has focused on whether in principle these quantum fluctuations could be exploited to construct a wormhole.

Although the energy density is the most important aspect of the matter and fields in the universe, other aspects of the content of spacetime (such as pressures and energy flows) are also important and must be specified in the mathematics of Einstein's theory. The mathematical object that includes all the information is called the "energy-momentum tensor," or alternatively, the "stress-energy tensor."

equation of state

The structure of a star is due to the combined action of gravity trying to pull the matter of a star inward, and forces within stellar material pressing outward. The relationship of that outward pressure to conditions, especially density, of the stellar material, is called the "equation of state of the material." For an ordinary star, the equation of state involves the physics of high-temperature gases and is well understood. For a **neutron star**, on the other hand, the equation of state depends on the detailed nature of nuclear forces and is inadequately understood. In his essay, Kip Thorne shows how observations of **gravitational waves** through the use of **laser interferometry** can help physicists to get a better understanding of the nuclear equation of state.

exotic material

Any form of matter and any field, like the electromagnetic field, has some **equation of state** relating its density (of mass, or of energy) to the pressure it exerts. For ordinary material, the magnitude of pressure within it is, in a sense, much smaller than its density. To construct a **wormhole** and a wormhole-based time machine requires material that is so far from ordinary that Kip Thorne and others call it "exotic." It is the subject of his ninth prediction and of the discussion toward the end of his essay.

expectation value

See **quantum theory**.

general relativity

"General relativity" is the name used for Einstein's theory of gravitation, which he introduced in 1915. It differs from Einstein's 1905 theory of spacetime, "special relativity," which applies only in the special circumstance that gravitational influences are absent. General relativity describes gravitational influences in terms of the curvature of spacetime; in special relativity, spacetime is flat (see curved and flat). In the Introduction (in the section "Why Is the Geometry of Spacetime 'Curved'?"), we showed that the idea of a curved spacetime has a very natural connection with gravitation.

Einstein's theory was the first, but is not the only, theory to use curved spacetime to describe gravitational influences. Many alternate theories have been proposed during this century. Most such theories differ only in the mathematical rules they give for how the content of spacetime (matter and energy) forces spacetime to curve. Mathematically, Einstein's general relativity has the simplest such rule among all these theories. This simplest of theories has so far passed every detailed experimental test of gravitation theory. Interestingly, Einstein modified his rule a little by introducing the **cosmological constant,** which is now also thought to play a role in the dynamics of the universe.

grandfather paradox

"Grandfather paradox" is the short name given to the contradictions that will occur if people can travel back in time and change events that have already occurred. As Stephen Hawking writes in his essay, "What happens to you if you go back in time and kill your grandfather before your father was conceived?" If this were allowed to happen, then you would not have been born, and you could not have traveled back in time to affect things that way. Physicists have done a great deal of work to demonstrate that time travel does not need to lead to such contradictions, provided all objects obey the deterministic laws of physics. (Hawking specifically excludes free will, which by definition is nondeterministic.) Igor Novikov describes in detail the way in which contradictions are avoided in some time-travel scenarios.

gravitational waves

See the section "Gravitational Waves" in the Introduction.

hyperspace

A two-dimensional world (a plane, a potato chip, the surface of Earth) can be mathematically described entirely in terms of its own internal geometry, the relationship of distances measured purely in that world. But it is a great intuitive aid to picture such geometries as surfaces in a flat three-dimensional world. Relativity theory deals with curved geometries of higher dimension, in particular with four-dimensional curved spacetime. It is sometimes useful to imagine these curved geometries of relativity theory as surfaces in some higher-dimensional flat (see **curved and flat**) space called hyperspace. Both Igor Novikov and Kip Thorne use this imagery in their essays. Although hyperspace is useful for intuitive pictures, it rarely enters into the mathematics of relativity. Typically, relativity research is done in terms of the internal geometry of four-dimensional spacetime, and relativists do not refer to a higher-dimensional hyperspace.

laser interferometry

When a wave signal due to a single source comes to us along two different paths, the separate contributions of the wave will combine in a complicated pattern of cancellations and reinforcements called "interference." The multicolored stripes on a thin layer of oil are due to the interference between the light waves reflected from the top and bottom of the oil layer; the position of the stripes depends on the thickness of the layer. In interferometry, the position of the stripes is used in this way as a way to measure thicknesses. This technique can be extended to the measurement of large "thicknesses" such as the distance between two mirrors separated by several kilometers, but only the light produced by lasers is pure enough to form stripes for such distances. Because a laser interferometer allows exquisite precision in measuring the separation of two objects, it is well suited to sense the oscillation in separation that will occur when a passing gravitational wave acts on the two distant masses.

metric

The beginning of the Introduction focuses on the idea of quantifying the distances separating points in space or in spacetime. The expression "metric" refers to the formula that gives the distance. To specify the metric is to specify the geometry. Thus, when Stephen Hawking writes, in his essay, about "quantum fluctuations of the metric," he is referring to the possibility of a **quantum theory** for the geometry of spacetime. Just as quantum effects impose uncertainty on the orbit of an electron, Stephen Hawking tells us that it will impose uncertainty in the very geometry of spacetime itself.

naked singularity

See **cosmic censorship.**

neutron star

Most of the mass of an atom is in its nucleus, but the size of a normal atom is determined by the atom's electrons, which occupy a space much larger than the size of the nucleus. In some stars, gravitational forces can be strong enough that the (negatively charged) electrons of atoms in the stellar material are compressed into the nuclei, joining with (positively charged) protons to become neutrons. Almost the entire star then consists of only neutrons packed extremely tightly together. This extremely dense "nuclear matter," unlike ordinary matter, has an **equation of state** governed primarily by nuclear forces.

The detailed physics of gravity and nuclear forces constrains neutron stars to be no more than several times as massive as our Sun, although much, much smaller in size. A neutron star, in fact, is not much larger than the **black hole horizon** of a hole of the same mass. Both neutron stars and black holes are so compact that they can form binary pairs of mutually orbiting compact objects that generate relatively strong **gravitational waves.** In his essay, Kip Thorne discusses how, using **laser interferometry,** detection of gravitational waves from such systems can give physicists important information about nuclear forces and the nuclear equation of state.

nonlinearity

The term "linear," in connection with equations, theories, and physical interactions, is *not* meant to describe straight lines. Rather, it means in a broad sense that things can be added. Classical electrodynamics is an example of a theory of a linear interaction. If we calculate the electric field due to a first electric charge, and then the electric field due to a second charge, we can add the two fields we have just calculated in order to find the field produced by the two charges together.

In Einstein's **general relativity,** gravity does not work that way. The gravitational force due to two bodies is not simply the sum of the forces due to each. Interactions in which simple summation does not give the correct answer, such as those involving gravity, are said to be nonlinear. Much of the technical difficulty of working with Einstein's theory has to do with the nonlinear character of the theory.

numerical relativity

The mathematics of Einstein's **general relativity** takes the form of a set of very difficult equations for the details of the **metric** that governs the geometry of spacetime. Succinct mathematical solutions to these equations can be found only for very simple configurations, such as spherically symmetric spacetime. Since the 1970s, much effort has gone into "numerical relativity," the solution of Einstein's equations using large computers. Eventually, numerical relativity will produce solutions for realistically complicated astrophysical configurations, but progress with these computer solutions is extremely difficult, partly due to the **nonlinearity** in Einstein's equations. Kip Thorne, in his essay, points out that physicists hoping to detect gravitational waves from the inspiral of pairs of compact objects are counting on numerical relativity to provide the theoretical details of those waves.

Planck time

At present there is no theory that combines the curved spacetime gravity of Einstein with the principles of **quantum theory,** yet it is possible to make fairly secure statements about some features of a full theory. One such statement, and an important one, has to do with the "Planck units" (also called "Planck-Wheeler units")—the characteristic energy, time, and distance of interactions that combine curved spacetime gravity and quantum theory on an equal footing. Physicists reason that these units can be constructed from the fundamental constants of physical theory: the light speed built into special relativity, the Planck constant of quantum theory, and the universal constant of gravitation theory. The Planck time turns out to be $\sqrt{Gh/c^5}$, which works out to be around 10^{-34} second.

quantum fields

See **quantum theory.**

quantum fluctuations

See **quantum theory.**

quantum theory

Quantum theory, along with relativity, stands as one of the two great advances of theoretical physics in the twentieth century. "Quantum theory" is a general term used for theories in which the laws of physics predict only the probabilities of various outcomes of experiments. In a quantum theory, if one knows everything possible about a physical system today, there is no guarantee that one can predict its condition tomorrow with complete precision. One can still predict the likely outcome of an experiment, which is called its "expectation value," but any single actual experiment might produce a different value. The term "quantum fluctuation" is used to suggest the way in which the values of physical quantities vary in quantum theories.

Quantum theories contrast sharply with "classical" theories, deterministic theories, in which the precise future state of a physical system can in principle be predicted. Classical theories, like Newton's mechanics and Maxwell's electrodynamics, turned out to be inadequate to describe the structure of atoms, the behavior of atoms in molecules, and the behavior of light under certain circumstance.

There are various quantum theories. Originally physicists described atoms using quantum theory, but described the electromagnetic fields in atoms with fully deterministic laws. Later, physicists devised successful theories of "quantum fields," in which fields such as electromagnetism had quantum fluctuations. It is interesting that fluctuations of fields occur even when there is no field present according to classical expectations. These are called "vacuum fluctuations" and are well established aspects of quantum field theory. Kip Thorne points out near the end of his essay that vacuum fluctuations may play a crucial role in preventing the existence of **wormholes.**

At present, physicists use theories that treat spacetime itself as if it were classical. These hybrid theories (classical spacetime, quantum content of spacetime) are called "semiclassical" theories. A long-standing challenge of theoretical physics is to find a "full quantum theory" that describes the fluctuations of spacetime, or of the **metric,** as well as everything in it. Such a theory, which would replace **general relativity,** is sometimes called "quantum gravity".

A feature common to all quantum theories is that quantum fluctuations follow certain laws. The most important law is the "uncertainty principle," referred to by Stephen Hawking. According to the uncertainty principle, measurable quantities can be arranged in pairs that have correlated fluctuations. If the nature of the physical system makes one quantity of the pair very well determined, then the other member of the pair must be very ill determined; it must fluctuate wildly. The location and speed of a particle form one such pair. In a quantum gravity theory such a pair would be the location of an event in spacetime and the energy associated with that event. Some physicists believe that quantum fluctuations of spacetime near a **singularity** will "spread out" the classically predicted infinities and restore predictability (a limited, quantum predictability) to spacetime. (See **cosmic censorship.**)

semiclassical

See **quantum theory.**

signal-to-noise ratio

In any experiments or measurements or detections, scientists look for some "signal" containing the information they are after. Along with the signal, there is always an unwanted additional component caused, for example, by the imperfection of the detection equipment. This unwanted component is generically called "noise" though it is seldom related to sound, and the relative importance of the signal that is wanted, to the noise that is not, is called the "signal-to-noise ratio," or often simply "signal-to-noise."

In everyday communication technology, such as radio transmissions, the noise is often a small, almost unnoticeable part of what is received. The opposite is true of many scientific experiments. They are at the forefront of what is possible, and therefore by their very nature they deal with signals that are not strong compared with the noise that accompanies them. Gravitational-wave signals will be extraordinarily weak, so the question of the information that can be wrung from them largely depends on how successful scientists will be in extracting a small signal from much noise.

In his essay, Kip Thorne discusses the gravitational-wave information from the inspiral of a small black hole into a much bigger black hole. His predictions are based on expectations of the signal-to-noise capabilities in the years 2010 to 2015 with the space-based, gravitational-wave observatory called LISA (Laser Interferometer Space Antenna).

singularity

Physical theories make predictions about the size of quantities, and sometimes an infinite size is predicted. An example is the gravitational contraction of a sphere of matter with negligible pressure, something like a spherical ball of dust. According to Newton's classical theory of gravity, the gravitational pull inward acts without limit and compresses the body to a zero radius, and hence to an infinite density of matter. This infinite density is considered a "singularity," a deviation from the continuous finite behavior that we expect for physical quantities. The singular behavior of a spherical body in Newton's theory is easily explained away as a consequence of the unreasonable assumption that the body is precisely spherical.

In **general relativity,** Einstein's theory of the geometry of space-time, singularities generally correspond to an infinite curvature of spacetime (see **curved and flat** and **warpage**). Unlike Newtonian singularities, singularities in Einstein's theory are not easily dismissed as artifacts of unrealistic configurations. General relativity singularities form for wide ranges of conditions, and there are at least two astrophysical cases in which we can count on singularities. There is a singularity inside the **black hole horizon** whenever a black hole forms, and the big bang, the birth of the universe, is itself a singularity.

It is generally assumed that theoretical singularities in the space-time are due to the incompleteness of Einstein's theory, and that singularities will not show up in a more complete theory that combines Einstein's theory with **quantum theory.**

sum over histories

The Nobel Prize–winning Caltech physicist Richard Feynman (1918–1988) made many important contributions to **quantum theory,** especially to the theory of quantum fields. One of his most intriguing was to show that particles behave in quantum theory as if they somehow probe all possible spacetime paths between a starting and an ending event. In classical physics, only one path is taken, but in quantum physics there is some probability of any spacetime path (usually with the greatest probability going to the expected classical path). The different outcomes of an experiment in quantum theory are therefore viewed as different paths taken in spacetime. Feynman called this way of calculating quantum fluctuations the "sum over histories" It is an attractive way to formulate quantum gravitation, because it does not involve assumptions that spacetime is smooth and continuous on small scales.

surface of constant time

In spacetime, the expression "at a given instant of time" means all events with the same time. These events could have any spatial location whatever, and hence "at a given instant of time" is a three-dimensional set of a points, or a three dimensional surface, called a surface of constant time. (In ordinary space, three dimensions would be called a volume. In four-dimensional spacetime, the term "surface" is used for three-dimensional structures or two-dimensional structures.)

Stephen Hawking discusses the possibility of a civilization modifying the spacetime to the future of a surface of constant time S. This is a precise way of talking about that civilization modifying spacetime after a certain instant of time.

timelike and spacelike

Suppose two events occur at the same spatial location but at different times in some reference frame. It is shown in the Introduction (see the section "Spacetime Diagrams") that there cannot be a reference frame in which the events occur at the same time. Those events will have a time difference between them in any reference frame. Such events are said to have a timelike separation. By contrast, events that do occur at the same time in some reference frame are said to have a spacelike connection.

Any two events that a given physical particle experiences must have a timelike separation. Thus the **worldline** of such a physical object consists of points all of which have a timelike relationship to each other.

two-point function

Just as general relativity has problems with **singularities,** so, too, does **quantum theory,** although in a different way. In quantum field theory, the location of an event in spacetime and the energy associated with that event are a pair of "uncertainty principle" quantities. Because quantum fluctuations are larger in quantities that are defined on very small scales, they get infinitely large when we try to measure the properties of spacetime, or of physical fields, over infinitesimally short distances. Large fluctuations in energy would lead to conflicts, such as making small particles infinitely heavy. Scientists have found several equivalent ways to subtract away the effects of these fluctuations, leading to theories that are consistent with experiment. Subtracting them is harder in the curved spacetime of **general relativity,** but the danger of not doing so is greater, because the fluctuations would destroy the smoothness of spacetime itself. One of the methods, described by Stephen Hawking, examines the way these fluctuations occur at two nearby points and uses the prosaically named "two-point function" to remove the infinities, just as in flat spacetime. Hawking examines what happens when the curved spacetime contains **closed timelike curves,** a test that this method does not have to pass when it is used to remove singularities in flat spacetime!

Because quantum fluctuations contain more energy when they occur over short distances, it is possible to find a distance so small that the energy of the fluctuation is large enough to form a tiny black hole whose **black hole horizon** is the same size as the small distance. Physicists expect that it will not be possible for spacetime to be smooth over such short distances. Hawking speculates that this provides a natural "cutoff" in the fluctuations, that nature may not in fact produce singular fluctuations but rather limit them to this smallest size and largest energy.

vacuum fluctuation

See **quantum theory.**

warpage

When spacetime is not flat (see **curved and flat**), there is a way of quantifying how curved it is. In the case of a two-dimensional surface, curvature is quantified with two numbers, the maximum and minimum radii of curvature of the surface. The smaller these radii are, the more curved the surface is. The details are more complicated for higher dimensions, but the general picture of quantification remains valid. In discussing the future of gravitational wave astronomy, Kip Thorne uses "warpage" to refer to the general magnitude of curvature in the highly curved spacetime just outside a black hole, a region that can be probed with **gravitational waves.**

waveform

Waves (sound, electromagnetic, gravitational, and so forth) from natural sources are seldom simple oscillations of fixed amplitude and period. Rather, the complex details of the source produce complex details in the "waveform," the variation with time of the wave signal. An example of this is the complex waveform of sound from an orchestra, a waveform that contains information about many musical instruments. In his essay, Kip Thorne predicts that in the not-too-distant future, waveforms from **gravitational waves** will be detected by **laser interferometry** with such good **signal-to-noise ratio** that scientists will be able to learn much about the **warpage** of spacetime near a black hole, and about much more.

worldline

On a spacetime diagram (see "Spacetime Diagrams" in the Introduction), lines (not necessarily straight) are often drawn representing the continuous stream of events occurring to a physical object. Such a line is called the "worldline" of that object. The concept is so valuable that it is often used without reference to a specific spacetime diagram, to suggest the general idea of how the object is moving.

wormholes

In simple, flat, Euclidean space—the space that we intuitively believe we live in—there is only one shortest path between two places. Any wiggle introduced in that path makes it longer. Because Einstein's theory tells us that space is not flat, more interesting possibilities exist for the ways in which spatial locations are connected. It is possible that two (or more) distinct paths exist between two locations. Both paths could be "shortest" in the sense that any small wiggle makes the path longer. The two paths need not be the same length, and in fact one of the paths can be much shorter than the other. For such a situation, spacetime physicists call the shorter path a "wormhole." The places where this path becomes distinct are called the mouths of the wormhole.

A wormhole is a nontrivial structure in curved three-dimensional space, and cannot easily be visualized. What can be done is to use a two-dimensional example of the same kind of connection. This is done in the Introduction (see especially Figure 9). In his essay on traveling to the past, Igor Novikov describes a wormhole using a picture of two-dimensional wells whose gravitational wells connect to each other. In these three-dimensional pictures, a two-dimensional wormhole typically appears as if it is the boundary of a narrow tunnel, a hole that a worm might make in the ground.

The existence of a wormhole shortcut can allow travel between two places in a very short time, less than the time it takes light to travel between those two places by the standard route. In some sense, this is apparent faster-than-light travel. (It isn't really faster than light: a photon will go faster through the wormhole than any particle.) Although it is not at all obvious, this kind of faster-than-light travel makes it possible to return to an earlier time and to follow a **closed timelike curve.** In the Introduction, this is accomplished through the use of two wormholes moving with respect to each other. In Igor Novikov's essay, it is done through a combination of gravity and a single wormhole.

During the last decade, there has been very active research on whether the laws of physics allow wormholes to exist. The presently accepted answer "probably not" is one of the subjects of the essay by Stephen Hawking.

More details about wormholes and their connection to time travel can be found in Chapter 14 of *Black Holes and Time Warps: Einstein's Outrageous Legacy* by Kip S. Thorne (W. W. Norton, New York, 1994).

INDEX